Networks: A Very Short Introduction

VERY SHORT INTRODUCTIONS are for anyone wanting a stimulating and accessible way in to a new subject. They are written by experts, and have been published in more than 25 languages worldwide.

The series began in 1995 and now represents a wide variety of topics in history, philosophy, religion, science, and the humanities. The VSI library now contains more than 300 volumes — a Very Short Introduction to everything from ancient Egypt and Indian philosophy to conceptual art and cosmology — and will continue to grow in a variety of disciplines.

VERY SHORT INTRODUCTIONS AVAILABLE NOW:

ADVERTISING Winston Fletcher
AFRICAN HISTORY John Parker and Richard Rathbone
AGNOSTICISM Robin Le Poidevin
AMERICAN HISTORY Paul S. Boyer
AMERICAN IMMIGRATION David A. Gerber
AMERICAN POLITICAL PARTIES AND ELECTIONS L. Sandy Maisel
THE AMERICAN PRESIDENCY Charles O. Jones
ANAESTHESIA Aidan O'Donnell
ANARCHISM Colin Ward
ANCIENT EGYPT Ian Shaw
ANCIENT GREECE Paul Cartledge
ANCIENT PHILOSOPHY Julia Annas
ANCIENT WARFARE Harry Sidebottom
ANGELS David Albert Jones
ANGLICANISM Mark Chapman
THE ANGLO-SAXON AGE John Blair
THE ANIMAL KINGDOM Peter Holland
ANIMAL RIGHTS David DeGrazia
THE ANTARCTIC Klaus Dodds
ANTISEMITISM Steven Beller
ANXIETY Daniel Freeman and Jason Freeman
THE APOCRYPHAL GOSPELS Paul Foster
ARCHAEOLOGY Paul Bahn
ARCHITECTURE Andrew Ballantyne
ARISTOCRACY William Doyle
ARISTOTLE Jonathan Barnes
ART HISTORY Dana Arnold
ART THEORY Cynthia Freeland
ATHEISM Julian Baggini
AUGUSTINE Henry Chadwick
AUSTRALIA Kenneth Morgan
AUTISM Uta Frith
THE AZTECS Davíd Carrasco
BARTHES Jonathan Culler
BEAUTY Roger Scruton
BESTSELLERS John Sutherland
THE BIBLE John Riches
BIBLICAL ARCHAEOLOGY Eric H. Cline
BIOGRAPHY Hermione Lee
THE BLUES Elijah Wald
THE BOOK OF MORMON Terryl Givens
BORDERS Alexander C. Diener
THE BRAIN Michael O'Shea
BRITISH POLITICS Anthony Wright
BUDDHA Michael Carrithers
BUDDHISM Damien Keown
BUDDHIST ETHICS Damien Keown
CANCER Nicholas James
CAPITALISM James Fulcher
CATHOLICISM Gerald O'Collins
THE CELL Terence Allen and Graham Cowling
THE CELTS Barry Cunliffe
CHAOS Leonard Smith
CHILDREN'S LITERATURE Kimberley Reynolds

CHINESE LITERATURE Sabina Knight
CHOICE THEORY Michael Allingham
CHRISTIAN ART Beth Williamson
CHRISTIAN ETHICS D. Stephen Long
CHRISTIANITY Linda Woodhead
CITIZENSHIP Richard Bellamy
CIVIL ENGINEERING David Muir Wood
CLASSICAL MYTHOLOGY Helen Morales
CLASSICS Mary Beard and John Henderson
CLAUSEWITZ Michael Howard
THE COLD WAR Robert McMahon
COLONIAL LATIN AMERICAN LITERATURE Rolena Adorno
COMMUNISM Leslie Holmes
THE COMPUTER Darrel Ince
THE CONQUISTADORS Matthew Restall and Felipe Fernández-Armesto
CONSCIENCE Paul Strohm
CONSCIOUSNESS Susan Blackmore
CONTEMPORARY ART Julian Stallabrass
CONTINENTAL PHILOSOPHY Simon Critchley
COSMOLOGY Peter Coles
CRITICAL THEORY Stephen Eric Bronner
THE CRUSADES Christopher Tyerman
CRYPTOGRAPHY Fred Piper and Sean Murphy
THE CULTURAL REVOLUTION Richard Curt Kraus
DADA AND SURREALISM David Hopkins
DARWIN Jonathan Howard
THE DEAD SEA SCROLLS Timothy Lim
DEMOCRACY Bernard Crick
DERRIDA Simon Glendinning
DESCARTES Tom Sorell
DESERTS Nick Middleton
DESIGN John Heskett
DEVELOPMENTAL BIOLOGY Lewis Wolpert
THE DEVIL Darren Oldridge
DICTIONARIES Lynda Mugglestone
DINOSAURS David Norman
DIPLOMACY Joseph M. Siracusa
DOCUMENTARY FILM Patricia Aufderheide
DREAMING J. Allan Hobson
DRUGS Leslie Iversen
DRUIDS Barry Cunliffe
EARLY MUSIC Thomas Forrest Kelly
THE EARTH Martin Redfern
ECONOMICS Partha Dasgupta
EGYPTIAN MYTH Geraldine Pinch
EIGHTEENTH-CENTURY BRITAIN Paul Langford
THE ELEMENTS Philip Ball
EMOTION Dylan Evans
EMPIRE Stephen Howe
ENGELS Terrell Carver
ENGINEERING David Blockley
ENGLISH LITERATURE Jonathan Bate
ENVIRONMENTAL ECONOMICS Stephen Smith
EPIDEMIOLOGY Rodolfo Saracci
ETHICS Simon Blackburn
THE EUROPEAN UNION John Pinder and Simon Usherwood
EVOLUTION Brian and Deborah Charlesworth
EXISTENTIALISM Thomas Flynn
FASCISM Kevin Passmore
FASHION Rebecca Arnold
FEMINISM Margaret Walters
FILM Michael Wood
FILM MUSIC Kathryn Kalinak
THE FIRST WORLD WAR Michael Howard
FOLK MUSIC Mark Slobin
FORENSIC PSYCHOLOGY David Canter
FORENSIC SCIENCE Jim Fraser
FOSSILS Keith Thomson

FOUCAULT Gary Gutting
FREE SPEECH Nigel Warburton
FREE WILL Thomas Pink
FRENCH LITERATURE John D. Lyons
THE FRENCH REVOLUTION William Doyle
FREUD Anthony Storr
FUNDAMENTALISM Malise Ruthven
GALAXIES John Gribbin
GALILEO Stillman Drake
GAME THEORY Ken Binmore
GANDHI Bhikhu Parekh
GENIUS Andrew Robinson
GEOGRAPHY John Matthews and David Herbert
GEOPOLITICS Klaus Dodds
GERMAN LITERATURE Nicholas Boyle
GERMAN PHILOSOPHY Andrew Bowie
GLOBAL CATASTROPHES Bill McGuire
GLOBAL ECONOMIC HISTORY Robert C. Allen
GLOBAL WARMING Mark Maslin
GLOBALIZATION Manfred Steger
THE GOTHIC Nick Groom
GOVERNANCE Mark Bevir
THE GREAT DEPRESSION AND THE NEW DEAL Eric Rauchway
HABERMAS James Gordon Finlayson
HEGEL Peter Singer
HEIDEGGER Michael Inwood
HERODOTUS Jennifer T. Roberts
HIEROGLYPHS Penelope Wilson
HINDUISM Kim Knott
HISTORY John H. Arnold
THE HISTORY OF ASTRONOMY Michael Hoskin
THE HISTORY OF LIFE Michael Benton
THE HISTORY OF MATHEMATICS Jacqueline Stedall
THE HISTORY OF MEDICINE William Bynum
THE HISTORY OF TIME Leofranc Holford-Strevens
HIV/AIDS Alan Whiteside
HOBBES Richard Tuck
HUMAN EVOLUTION Bernard Wood
HUMAN RIGHTS Andrew Clapham
HUMANISM Stephen Law
HUME A. J. Ayer
IDEOLOGY Michael Freeden
INDIAN PHILOSOPHY Sue Hamilton
INFORMATION Luciano Floridi
INNOVATION Mark Dodgson and David Gann
INTELLIGENCE Ian J. Deary
INTERNATIONAL MIGRATION Khalid Koser
INTERNATIONAL RELATIONS Paul Wilkinson
ISLAM Malise Ruthven
ISLAMIC HISTORY Adam Silverstein
ITALIAN LITERATURE Peter Hainsworth and David Robey
JESUS Richard Bauckham
JOURNALISM Ian Hargreaves
JUDAISM Norman Solomon
JUNG Anthony Stevens
KABBALAH Joseph Dan
KAFKA Ritchie Robertson
KANT Roger Scruton
KEYNES Robert Skidelsky
KIERKEGAARD Patrick Gardiner
THE KORAN Michael Cook
LANDSCAPES AND GEOMORPHOLOGY Andrew Goudie and Heather Viles
LANGUAGES Stephen R. Anderson
LATE ANTIQUITY Gillian Clark
LAW Raymond Wacks
THE LAWS OF THERMODYNAMICS Peter Atkins
LEADERSHIP Keith Grint
LINCOLN Allen C. Guelzo
LINGUISTICS Peter Matthews
LITERARY THEORY Jonathan Culler

LOCKE John Dunn
LOGIC Graham Priest
MACHIAVELLI Quentin Skinner
MADNESS Andrew Scull
MAGIC Owen Davies
MAGNA CARTA Nicholas Vincent
MAGNETISM Stephen Blundell
THE MARQUIS DE SADE John Phillips
MARTIN LUTHER Scott H. Hendrix
MARX Peter Singer
MATHEMATICS Timothy Gowers
THE MEANING OF LIFE Terry Eagleton
MEDICAL ETHICS Tony Hope
MEDIEVAL BRITAIN John Gillingham and Ralph A. Griffiths
MEMORY Jonathan K. Foster
METAPHYSICS Stephen Mumford
MICHAEL FARADAY Frank A. J. L. James
MODERN ART David Cottington
MODERN CHINA Rana Mitter
MODERN FRANCE Vanessa R. Schwartz
MODERN IRELAND Senia Pašeta
MODERN JAPAN Christopher Goto-Jones
MODERN LATIN AMERICAN LITERATURE Roberto González Echevarría
MODERNISM Christopher Butler
MOLECULES Philip Ball
THE MONGOLS Morris Rossabi
MORMONISM Richard Lyman Bushman
MUHAMMAD Jonathan A. C. Brown
MULTICULTURALISM Ali Rattansi
MUSIC Nicholas Cook
MYTH Robert A. Segal
NATIONALISM Steven Grosby
NELSON MANDELA Elleke Boehmer
NEOLIBERALISM Manfred Steger and Ravi Roy
NETWORKS Guido Caldarelli and Michele Catanzaro
THE NEW TESTAMENT Luke Timothy Johnson
THE NEW TESTAMENT AS LITERATURE Kyle Keefer
NEWTON Robert Iliffe
NIETZSCHE Michael Tanner
NINETEENTH-CENTURY BRITAIN Christopher Harvie and H. C. G. Matthew
THE NORMAN CONQUEST George Garnett
NORTH AMERICAN INDIANS Theda Perdue and Michael D. Green
NORTHERN IRELAND Marc Mulholland
NOTHING Frank Close
NUCLEAR POWER Maxwell Irvine
NUCLEAR WEAPONS Joseph M. Siracusa
NUMBERS Peter M. Higgins
OBJECTIVITY Stephen Gaukroger
THE OLD TESTAMENT Michael D. Coogan
THE ORCHESTRA D. Kern Holoman
ORGANIZATIONS Mary Jo Hatch
PAGANISM Owen Davies
PARTICLE PHYSICS Frank Close
PAUL E. P. Sanders
PENTECOSTALISM William K. Kay
THE PERIODIC TABLE Eric R. Scerri
PHILOSOPHY Edward Craig
PHILOSOPHY OF LAW Raymond Wacks
PHILOSOPHY OF SCIENCE Samir Okasha
PHOTOGRAPHY Steve Edwards
PLAGUE Paul Slack
PLANETS David A. Rothery
PLANTS Timothy Walker
PLATO Julia Annas
POLITICAL PHILOSOPHY David Miller
POLITICS Kenneth Minogue
POSTCOLONIALISM Robert Young
POSTMODERNISM Christopher Butler

POSTSTRUCTURALISM Catherine Belsey
PREHISTORY Chris Gosden
PRESOCRATIC PHILOSOPHY Catherine Osborne
PRIVACY Raymond Wacks
PROBABILITY John Haigh
PROGRESSIVISM Walter Nugent
PROTESTANTISM Mark A. Noll
PSYCHIATRY Tom Burns
PSYCHOLOGY Gillian Butler and Freda McManus
PURITANISM Francis J. Bremer
THE QUAKERS Pink Dandelion
QUANTUM THEORY John Polkinghorne
RACISM Ali Rattansi
RADIOACTIVITY Claudio Tuniz
THE REAGAN REVOLUTION Gil Troy
REALITY Jan Westerhoff
THE REFORMATION Peter Marshall
RELATIVITY Russell Stannard
RELIGION IN AMERICA Timothy Beal
THE RENAISSANCE Jerry Brotton
RENAISSANCE ART Geraldine A. Johnson
RISK Baruch Fischhoff and John Kadvany
RIVERS Nick Middleton
ROBOTICS Alan Winfield
ROMAN BRITAIN Peter Salway
THE ROMAN EMPIRE Christopher Kelly
THE ROMAN REPUBLIC David M. Gwynn
ROMANTICISM Michael Ferber
ROUSSEAU Robert Wokler
RUSSELL A. C. Grayling
RUSSIAN HISTORY Geoffrey Hosking
RUSSIAN LITERATURE Catriona Kelly
THE RUSSIAN REVOLUTION S. A. Smith
SCHIZOPHRENIA Chris Frith and Eve Johnstone
SCHOPENHAUER Christopher Janaway
SCIENCE AND RELIGION Thomas Dixon
SCIENCE FICTION David Seed
THE SCIENTIFIC REVOLUTION Lawrence M. Principe
SCOTLAND Rab Houston
SEXUALITY Véronique Mottier
SHAKESPEARE Germaine Greer
SIKHISM Eleanor Nesbitt
SLEEP Steven W. Lockley and Russell G. Foster
SOCIAL AND CULTURAL ANTHROPOLOGY John Monaghan and Peter Just
SOCIALISM Michael Newman
SOCIOLOGY Steve Bruce
SOCRATES C. C. W. Taylor
THE SOVIET UNION Stephen Lovell
THE SPANISH CIVIL WAR Helen Graham
SPANISH LITERATURE Jo Labanyi
SPINOZA Roger Scruton
SPIRITUALITY Philip Sheldrake
STARS Andrew King
STATISTICS David J. Hand
STEM CELLS Jonathan Slack
STUART BRITAIN John Morrill
SUPERCONDUCTIVITY Stephen Blundell
TERRORISM Charles Townshend
THEOLOGY David F. Ford
THOMAS AQUINAS Fergus Kerr
TOCQUEVILLE Harvey C. Mansfield
TRAGEDY Adrian Poole
TRUST Katherine Hawley
THE TUDORS John Guy
TWENTIETH-CENTURY BRITAIN Kenneth O. Morgan
THE UNITED NATIONS Jussi M. Hanhimäki
THE U.S. CONGRESS Donald A. Ritchie

THE U.S. SUPREME COURT
Linda Greenhouse
UTOPIANISM Lyman Tower Sargent
THE VIKINGS Julian Richards
VIRUSES Dorothy H. Crawford
WITCHCRAFT Malcolm Gaskill
WITTGENSTEIN A. C. Grayling
WORLD MUSIC Philip Bohlman
THE WORLD TRADE ORGANIZATION Amrita Narlikar
WRITING AND SCRIPT
Andrew Robinson

Available soon:

WORK Stephen Fineman
SPIRITUALITY
Philip Sheldrake
MARTYRDOM Jolyon Mitchell
RASTAFARI Ennis B. Edmonds
COMEDY Matthew Bevis

Guido Caldarelli and Michele Catanzaro

NETWORKS

A Very Short Introduction

OXFORD
UNIVERSITY PRESS

Great Clarendon Street, Oxford, OX2 6DP,
United Kingdom

Oxford University Press is a department of the University of Oxford.
It furthers the University's objective of excellence in research, scholarship,
and education by publishing worldwide. Oxford is a registered trade mark of
Oxford University Press in the UK and in certain other countries

First Edition published in 2012
Impression: 7

British Library Cataloguing in Publication Data
Data available

Library of Congress Cataloging in Publication Data
Data available

ISBN 978-0-19-958807-7

Printed in Great Britain by
Ashford Colour Press Ltd, Gosport, Hampshire

To my family
G. C.

To Anna
M. C.

Contents

List of illustrations

Chapter 1
A network point of view on the world

Networks are present in the everyday life of many people. On a typical day, we check emails, update social network profiles, make mobile phone calls, use public transportation, take planes, transfer money and goods, or start new personal and professional relations... In all these cases—consciously or not—we are using networks and their properties. Similarly, networks appear in important global phenomena. Financial crises generate domino effects in the web of connections between banks and companies. Pandemics—like avian flu, SARS, or swine flu—spread in the airport network. Climate change can alter the network of relations between species in ecosystems. Terrorism and war target the infrastructure grids of a country. Large-scale blackouts take place in power grids. Computer viruses diffuse in the Internet. Governments and companies can track people's identity through their social networks and other digital communication tools. Finally, the various applications of genetics depend on the knowledge of the genetic regulatory networks that operate within the cell.

In all these situations we deal with a large set of different elements (individuals, companies, airports, species, power stations, computers, genes...), connected through a disordered pattern of many different interactions: that is, they all have an underlying network structure. Often, this hidden network is the key to understanding those situations. A good example is the collapse of cod population in the north-western Atlantic in the eighties. At the time, the shortage of cod generated a massive economic crisis in the Canadian fishery industry. Canadian stakeholders asked for more expeditions to hunt seals, maintaining that controlling these predators of cod would help stop the collapse. Many seals were killed during the nineties, but the cod population did not recover. In the meantime, ecologists studied the different food chains that connected cod and seals. By the end of the decade, they drew a complete map (Figure 1) where a lot of different chains were found to connect the two species. In the light of this intricate picture, hunting for predators of cod does not necessarily help the fish. For example, seals predate about 150 species, and several of them are predators of cod: thus, reducing the population of seals can end up increasing the pressure of other predators on cod.

Ecosystems are complex webs of species: it is crucial to take into account this underlying network structure if we want to understand and manage them. Similar caution must be taken with several other systems, all based on a networked architecture. For example, the development of an infectious disease like AIDS is strongly influenced by the pattern of non-protected sexual relations within a population. Similarly, liquidity shocks depend on the interwoven network of money exchange between banks.

All the examples above are instances of the so-called *emergent phenomena*. That is some collective behaviour that cannot be predicted by looking at the single elements forming the system. Usually, systems that display these phenomena are dubbed *complex systems*. For example, a single ant is a relatively awkward animal, but many ants together are capable of activities as complex as building large anthills or storing large quantities of food. In human societies, social order arises from the combination of autonomous

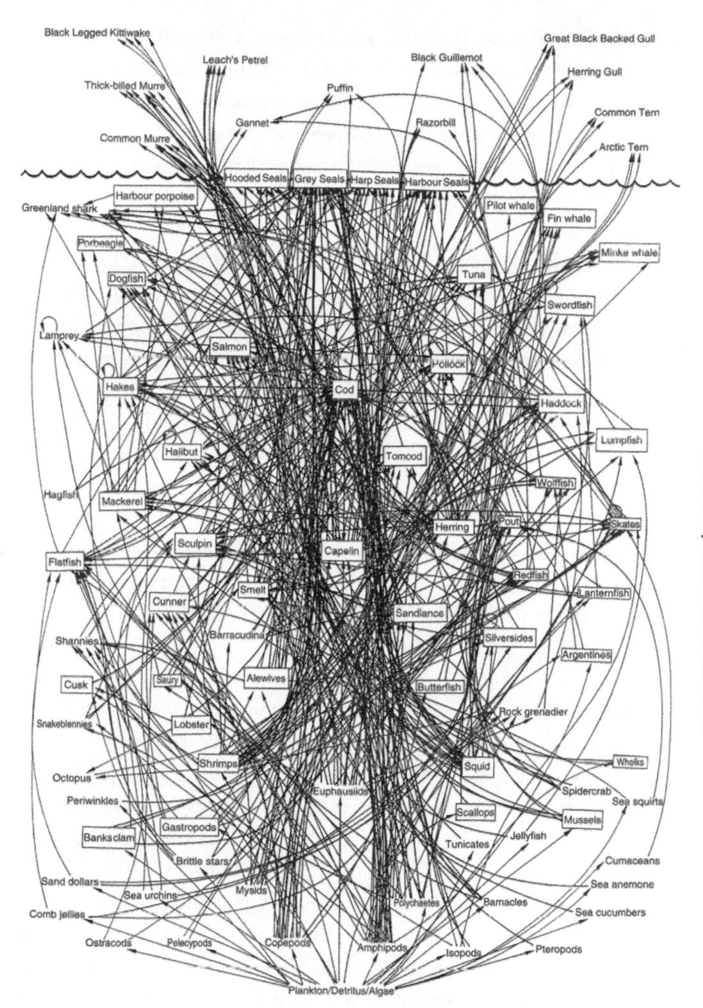

1. A partial food web for the 'Scotian Shelf' in the north-west Atlantic off eastern Canada. Arrows go from the predator species to the prey species

individuals, often with conflicting interests, that still end up performing tasks that nobody could do on their own. Similarly, a living organism arises from the interaction of its parts; and the extraordinary resilience of the Internet to errors, attacks, and traffic peaks is a performance of the network as a whole rather than the result of the action of individual machines.

Networks, with their emphasis on the interactions, are the key to understanding many of these phenomena. Imagine two football teams whose players have very similar skills, and yet the two teams perform very differently: probably this difference depends on how good or bad the interactions are between the players on the pitch. Similarly, a single player can be good in his league team and bad in his national team because of the different positions he has with respect to the other players in the two groups. The performance of a team is a kind of emergent phenomenon, one that does depends not only on the quality of the single players or on the sum of their individual skills, but also on the network of interactions between them. Many emergent phenomena rely crucially on the structure of the underlying networks.

The network approach focuses all the attention on the global structure of the interactions within a system. The detailed properties of each element on its own are simply ignored. Consequently, systems as different as a computer network, an ecosystem, or a social group are all described by the same tool: a *graph*, that is, a bare architecture of nodes bounded by connections. This approach was originally developed in mathematics by Leonard Euler and later spread to a wide range of disciplines, including sociology, which has deployed it widely, and more recently physics, engineering, computer science, biology, and many others.

Representing widely different systems with the same tool can only be done by a high level of abstraction. What is lost in the specific description of the details is gained in the form of *universality*—that is, thinking about very different systems as if they were different realizations of the same theoretical structure. In this respect, the spread of a computer virus can be similar to flu; hacking a router can have the same effect as the extinction of a species in an ecosystem; and the growth of the World Wide Web (WWW) can be set alongside the increase in scientific literature.

This line of reasoning provides many insights. For instance, representing a system as a graph allows us to perceive large-scale structures that encompass apparently unrelated elements. In 2003, a trivial accident in the Swiss electric grid triggered a large-scale blackout affecting Sicily, 1,000 kilometres away. Focusing on the network structure allows us to see that faraway elements end up being strongly connected, through incredibly short paths of relation or communication. The current observation that two individuals geographically and socially apart—such as a rainforest inhabitant and a manager in the City of London—are connected by only 'six degrees of separation' is not far from reality, and it can be explained in terms of the network structure of social relations.

The network approach also sheds light on another important feature: the fact that certain systems that grow without external control are still capable of spontaneously developing an internal order. Cells or ecosystems are not 'designed' but nevertheless work in a robust way. Similarly, social groups and trends arise from an immense variety of different pressures and motivations but still display clear and definite shapes. The Internet and the WWW boomed without the presence of any regulating authority and were promoted by an enormous variety of unrelated agents: however, they usually work in a coherent and efficient way. All these are *self-organized processes*, i.e. phenomena in which order and organization are not the result of an external intervention or global blueprint but the outcome of local mechanisms or tendencies, iterated along thousands of interactions. Network models are able to describe in a clear and natural way how self-organization arises in many systems. As well, networks allow us to better understand dynamical processes such as the rapid spread of computer viruses, the large scale of pandemics, the sudden collapse of infrastructures, and the bursts of social phobias or music trends.

In the study of complex, emergent, and self-organized systems (the modern science of complexity), networks are becoming increasingly important as a universal mathematical framework, especially when massive amounts of data are involved. This is typically the case of individuals accumulating queries in search engines, updates in social websites, payments online, credit card data, financial transactions, GPS positions from mobile phones, etc. In all these situations networks are crucial instruments to sort out and organize these data, connecting individuals, products, news, etc. to each other. Similarly, molecular biology relies more and more on computational strategies to find order in the large amounts of data it produces. The same happens in many other fields of science, technology, health, environment, and society. In all of these, networks are becoming the paradigm to uncover the hidden architecture of complexity.

Chapter 2
A fruitful approach

Crossing Euler's bridges

In the Russian city of Kaliningrad, the island of Kneiphof stands in the Pregel River. Three centuries ago, the city was in Prussia, it was called Königsberg, and at that time seven bridges connected the island with the rest of the city (Figure 2 top). A riddle was popular in the town: was it possible to walk across the seven bridges, without crossing any of them twice? Nobody had ever been able to do this successfully. On the other hand, no formal proof was available about the feasibility of such a walk. The solution came from one of the most famous mathematicians of all times. In 1736, Leonard Euler drew a map of Königsberg in an unusual fashion. He represented the portions of mainland and the island as dots, and the bridges as lines, connecting the dots with each other (Figure 2 bottom).

When we cast the problem in this form, things become easier. By showing the network out of the city, Euler proved that the walk was impossible. His explanation is based on the following observation: for such a walk to be feasible, all the dots along the path must have an even number of connections. This is because every time one enters any part of the city via a bridge, one must leave via a different bridge. In general, each area must have an even number of bridges, e.g. 2, 4, 6. Only the start points and end points of the walk can have an odd number of links: at the start point, there can be

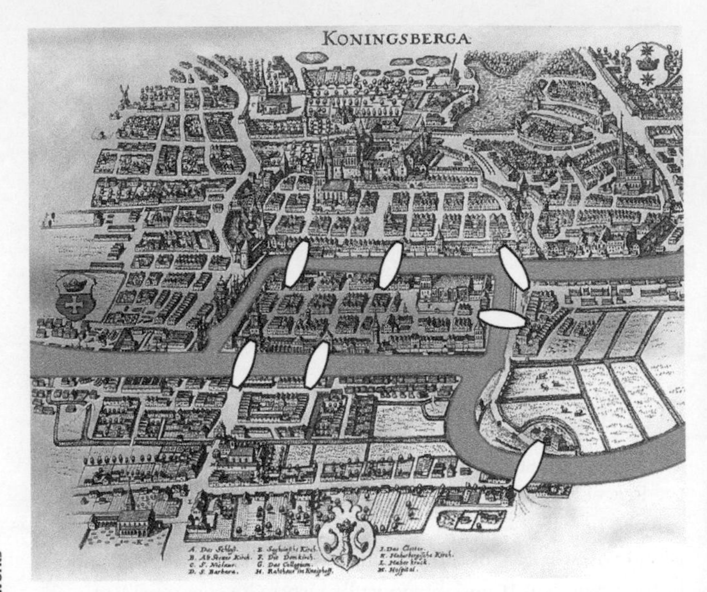

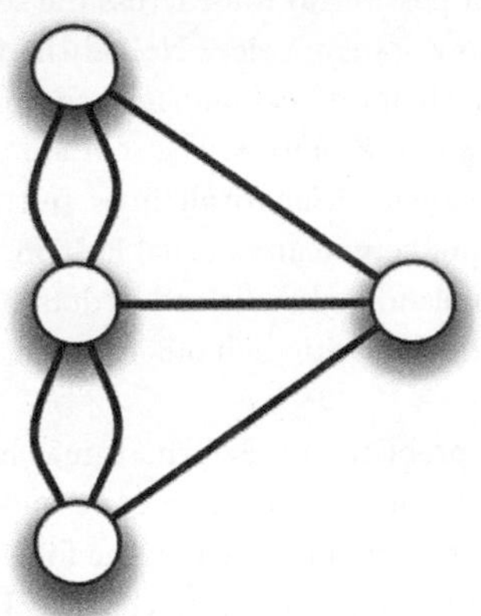

2. An engraving of Königsberg (top), now Kaliningrad, showing the Königsberg bridges riddle, represented by Leonard Euler as a graph (bottom)

only one bridge; and the same holds for the end point. Unfortunately, the graph of Königsberg has all the vertices with an odd number of links. Therefore it is impossible to walk only once over all the bridges.

Such a simpilified mathematical map of Königsberg is the first example of a *graph*. Mathematicians call the dots and lines forming it (respectively) *vertices* and *edges*. Nowadays, Euler is credited for having started a whole branch of mathematics, built on graph analysis. His intuition can be considered the first, foundational moment of network science. After him, many mathematicians studied the formal properties of networks, while scientists applied them to a wide range of problems: electrical circuits by G. R. Kirchhoff in 1845, isomers of organic components by A. Cayley in 1857, the mathematical concept known as the 'Hamiltonian cycle' by W. R. Hamilton in 1858, etc. One famous application is the 'map colouring problem', proposed in the mid-19th century. At that time, geographers were sorting out the minimum number of colours needed to draw maps, where adjacent countries have different colours. This was more than a theoretical problem: given the large number of countries and the reduced quantity of different inks available in the print industry, it was essential to be able to use the minimum number of colours. Empirically, three colours were not enough while four seemed to work well. The first proof that the solution was indeed four was obtained only in 1976. It is based on representing a map as a graph, where nodes are the countries and an edge is drawn between two of them if they share a border.

Runaway girls, Australian peoples, and Chicago workers

In the autumn of 1932, fourteen girls ran away in a period of just two weeks from Hudson School for Girls, in New York State. This was much more than the usual rate of runaways. School managers then decided to look into the individual personalities of the girls, to understand this phenomenon. Since no striking evidence of particular differences in the girls' personalities was found, the psychiatrist Jacob Moreno proposed a completely different explanation. He suggested that this large number was triggered by the position of runaways in the girls' social network. Moreno, together with Helen Jennings, mapped the social ties between the students, using *sociometry*, a technique for identifying relations between individuals. They found that these ties were the main channel through which the idea of running away spread among the girls. The position of an individual in the friendship network was crucial for replicating the behaviour of the runaway girls.

Moreno was one of the first researchers to apply the idea of network to society. After Euler's intuition, his work is a second and crucial step in network science foundations, starting one of the most important lines of network science: the analysis of social networks. Thirty years later, anthropologists applied a similar approach to kinship relations in tribes such as the Arunda people in Australia. In this case, the connections were drawn between individuals who were relatives. Researchers found that the resulting diagrams responded to elegant mathematical structures. These and other results suggested that neat social structures, or even universal laws, could be found underneath the disorder of the human social world. Since then, social sciences have widely deployed the concept of network to represent social structure. Many other studies followed these seminal ones, focusing on the circles of women in southern America (by A. Davis, B. B. Gardner, and M. R. Gardner in 1941), groups of factory workers in Chicago (by E. Mayo in 1939), friendship between schoolchildren (by A. Rapoport in 1961), and the relations between drug addicts in the city of Colorado Springs (by Richard B. Rothenberg et al. in 1995), among others.

Random connections

The third important moment in the foundation of network science came with the publication of a set of papers by the two mathematicians Paul Erdős and Alfréd Rényi, in 1959–61. Erdős, one of the most important mathematicians of the 20th century, has been described as 'a man who loved only numbers'. Partly wrong: he loved graphs too. The two theorists studied a mathematical model representing a graph where vertices are connected to each other completely at random. This model, lately known as Random Graph, was first proposed in 1951, in a paper by mathematics researchers Ray Salomonoff and Anatol Rapoport.

The random graph is a very simplified model, and its properties are very different from those of real networks. For example, randomness and chance could indeed play an important role in meeting new friends, but the formation of friendship networks is certainly related to many other factors, such as social class, common languages, affinity etc. However, the random graph model is very important because it quantifies the properties of a totally random network. Random graphs can be used as a benchmark, or *null case*, for any real network. This means that a random graph can be used in comparison to a real-world network, to understand how much chance has shaped the latter, and to what extent other criteria have played a role.

The simplest recipe for building a random graph is the following. We take all the possible pairs of vertices. For each pair, we toss a coin: if the result is heads, we draw a link; otherwise we pass to the next pair, until all the pairs are finished (this means drawing the link with a probability $p = \frac{1}{2}$, but we may use whatever value of p). Typically, the creation of the graph and its study are not done manually. Scientists use computer programs and draw the resulting network on paper or on computer screen. However, this becomes increasing difficult when the network is large. Moreover, it is hard to study an interwoven structure just by visual inspection. A better and quantitative insight comes from studying the graph as an abstract object, by means of mathematical tools. Computers can also help: simulation allows us to build, within the 'mind' of a computer, a faithful realization of the model, and then make measurements of it as if it were a real object. If we want to compare the abstract model with a real-world network, now we just need to compare the measures in both cases.

Since its introduction in the sixties, the random graph model has become one of the most successful mathematical models, despite its loose connection with reality. Nowadays it is a benchmark of comparison for all networks, since any deviations from this model suggest the presence of some kind of structure, order, regularity, and non-randomness in many real-world networks.

Nets to fish for information

In the wake of the terrorist attacks in New York in 2001, in Madrid in 2004, and in London in 2005, several governments have proposed, as an anti-terrorist measure, the storage of electronic traffic data. In this proposal, years of phone calls and emails between citizens would be recorded for safety purposes. The content of the messages would not be stored: it would be enough to record only senders and recipients (and sometimes time and place of communication). As police forces know, even this simple map of who is connected with whom is a powerful instrument in tracking people's activity. Indeed from a sketch of phone calls, we can deduce habits, circles of friends, and various other data about a person.

This is a very practical example—and a very debated one—of the basic approach of network science. A complex system is represented as a graph—a set of equal elements connected by equal interactions—ignoring the detailed features of its constituents, and the specific nature of their relations. This procedure may seem too drastic, but still it allows us to capture more information than expected at first sight. A proof of the effectiveness of this approach is the friendship advice system contained in many online social networks, such as Facebook or LinkedIn. The idea is simple: you are likely to know the friends of your friends. As simple as it is, it works most of the times. This method, called *educated guess,* is also behind the systems that suggest books or other items in online shops. The software of these companies exploits networks of goods associated with each consumer. This is why commercial companies store a great deal of electronic data, including email messages and online social network data: they know that this information is extremely useful.

The basic approach of network science can be applied to a broad set of systems. For example, an ecosystem composed of hundreds of species, where each one extracts energy from others by means of different strategies of predation. In the network approach, this is represented by a set of identical vertices connected by a set of identical edges. The same strategy is applied to systems ranging from the Internet (hundreds of thousands of computers, routers, exchange points, etc., connected by telephone lines, fibre optic cables, satellite communications, etc.) to a human population (a huge number of actors with different purposes and roles, connected by a variety of relations). While the network approach eliminates many of the individual features of the phenomenon considered, it still maintains some of its specific features. Namely, it does not alter the size of the system—i.e. the number of its elements—or the pattern of interaction—i.e. the specific set of connections between elements. Such a simplified model is nevertheless enough to capture the properties of the system.

From individuals to groups

There are two possible approaches when dealing with a situation where many different elements interact in different ways. In the first approach, we identify the basic constituents and interactions between them. By studying each element on its own, we can then deduce the behaviour of the system as the sum of its individual elements. For example, ecologists can describe the features of an ecosystem by listing prey and predators of every single species. Computer scientists describe a network of computers by focusing on the features and protocols of each different machine. Psychologists study social relations by describing the behaviour of each social actor in his or her circle.

A second strategy, different from the first one, consists in putting many elements together into a few homogeneous groups. For example, sociologists and political scientists usually split society into social classes, genders, levels of education, ethnicities, nations, etc. Similarly, epidemiologists often separate the population into a limited set of 'compartments': healthy individuals, infected, immunized, etc. Ecologists can also use this approach by aggregating into groups (*trophic species*) all the species that have similar roles in a foodweb.

The network approach tries to complement these two points of view. Many phenomena are impossible to explain if one focuses only on the behaviour of individual elements. For example, the dynamics of a species population within an ecosystem do not depend on the features of that species alone: the global network of prey–predator interactions must be taken into account. On the other hand, focusing on big classes of elements may not be useful either. For example, the political phenomena taking place in a country are hardly the outcome of a pre-existing national identity but rather of the specific pattern of social relations within that country. The network approach is somewhere between the description by individual elements and the description by big groups, bridging the two of them. In a certain sense, networks try to explain how a set of isolated elements are transformed, through a pattern of interactions, into groups and communities. In all cases where this pattern is relevant, the network approach provides essential insights.

Geography versus 'netography'

At the beginning of the 20th century, London's underground train service (the Tube) became so intricate that more and more complicated maps had to be issued from time to time, in order to orient the travellers. In 1931, after many attempts, Henry Beck, an employee of the company, changed the criteria for drawing the chart. Instead of embedding the lines on top of an actual map of London, Beck placed them in an abstract space (Figure 3).

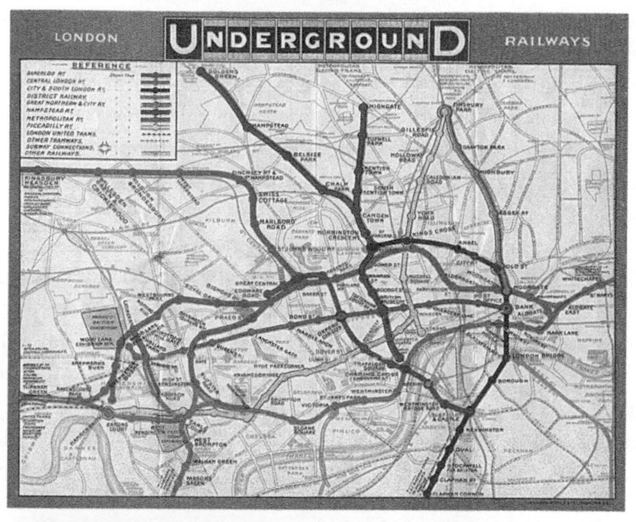

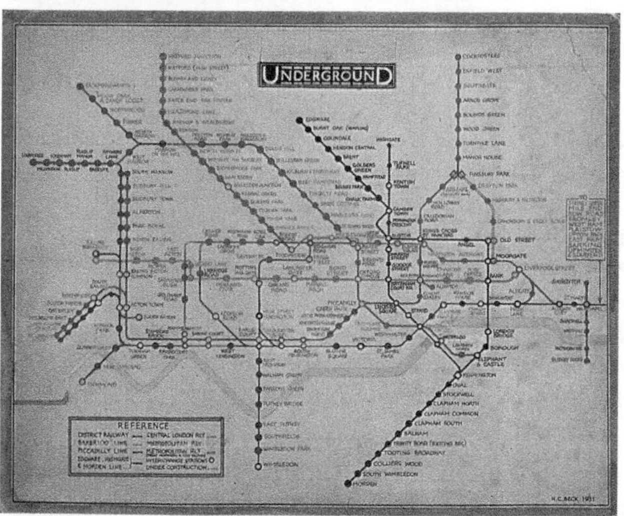

3. A 'metric' representation of the London Tube (top) versus a 'topological' one (bottom). Even though in the latter the actual positions and relative distances of the stations are not indicated, it is a more useful mind map of the service

Stations were represented by well-spaced dots. Tube connections became straight lines with neat angles of 45 or 90 degrees. This map has little to do with the real positions and distances of stations, but it is much clearer and more useful for the passengers. Those travelling on the Tube network are not interested in its geographic features: the information about the sequence of stations and the intersection of Tube lines is enough.

Henry Beck's London's Tube map is basically a graph. His solution to the mapping problem exploited a basic feature of the network approach: in networks, *topology* is more important than *metrics*. That is, what is connected to what is more important than how far apart two things are: in the other words, the physical geography is less important than the 'netography' of the graph. The difference between these two concepts is shown in Figure 4. The three images represented in the picture are different from a metric point of view. That is, the positions of nodes in space and the lengths of links are different. However, from the topological point of view they are identical: they are just three different representations of the same graph. In the network representation, the connections between the elements of a system are much more important than their specific positions in space and their relative distances.

The focus on topology is one of its biggest strengths of the network approach, useful whenever topology is more relevant than metrics. For example, an email sent from New York reaches an office in London in the same time as one sent from the office next door. Even on the Internet, a material infrastructure embedded in geographical space, the pattern of the connections is more

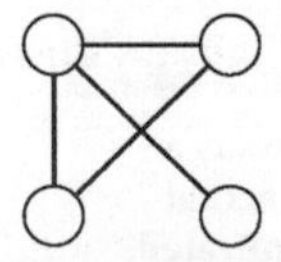
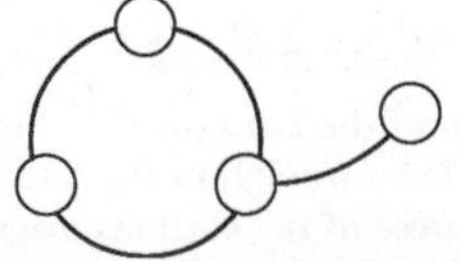
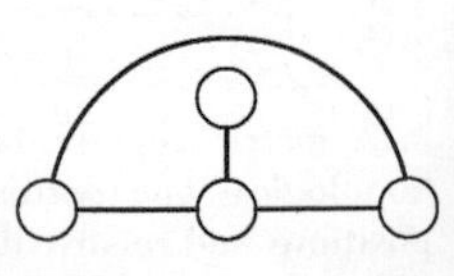

4. Three different representations of the same graph

important than the physical distance. In social networks, the relevance of topology means that *social structure* matters. Lionel Messi is nowadays one of the best football players in the world. However, his performance differs according to which team he is playing in (either Argentina or FC Barcelona). Some social scientists have argued that the network of Messi's relations with other players in Argentina is different from that in FC Barcelona. According to their research, this results in the player carrying a heavier 'weight' in the former case: this may explain, at least in part, the difference in his performance. Similar phenomena appear also in more complicated social 'games', where an individual's outcome can be strongly influenced by his or her position in a network of relations.

Chains, grids, and networks

The network approach reduces complex systems to a bare architecture of nodes and links. This is a substantial simplification, but still the resulting graph may not be so easy to interpret: this is the case with the tricky illustration shown in Figure 4. Even a graph as simple as an innocent chain of nodes can be a rather complicated object to handle. A chain may represent, for example, a fire brigade moving a bucket of water; or a food chain of species, in which the first predates the second, which predates the third, and so on; or a business-to-business supply structure: a set of companies in which each one supplies the next one.

Imagine a production chain of five companies (1, 2, 3, 4, and 5). Along this chain, any of them can make a deal with either of its two neighbours. The rule is that each company can close only one contract: for example, if 3 closes a deal with 2, it cannot have arrangements with 4. Given this simple structure and rule, it turns out that nodes 1 and 5 have less bargaining power, since they have fewer alternatives. This makes nodes 2 and 4 stronger, and (unexpectedly) it weakens node 3. Indeed, node 3 has only strong nodes to deal with, and therefore it ends up having less convenient deals. Something as simple as a linear sequence of nodes does indeed yield a rather complex landscape. This example shows what sociologists call an *exclusion mechanism*. Far from being a theoretical situation, this is commonly experienced in economics, when the establishment of a commercial relation between two parts excludes a third node.

To further complicate the question, one has to take into account that real-world systems are rarely as simple as a chain. In the past, scientists have represented complex systems through regular grids or *lattices*, instead of using graphs. These objects are composed of many elements—representing people, animals, computers, etc.—usually arranged along a regular pattern of connections, like pieces on a chessboard connecting only with their four neighbours. This regular structure makes systems much easier to handle by mathematical calculation and by computer simulation than when using a graph.

The lattice choice, albeit simpler with respect to a graph, introduces nevertheless a strong limitation. In fact, a lattice is suitable only for representing systems carefully designed or subject to strong constraints. These might be, for example, the arrays of processors in a computing cluster or verbal communications between chain workers in a noisy workplace. In lattices, every node is linked to a fixed number of nearest neighbours, while in the vast majority of real-world cases, links connect a variable number of elements, no matter whether close to or far away from each other. The ability to capture this disorder is one of the great advantages of the network approach.

Much of this disorder is encoded in a crucial quantity: the *degree*, that is, the number of edges attached to each node. If a node is a web page, the degree represents the number of links it receives from other pages. If a node is a species, the degree is the number of species it depends on for feeding. If a node is an individual, the degree is the number of acquaintances. This circle can be related to what sociologist Peter Marsden has called a *core discussion network*: the set of people (friends, partners, family members, current and past schoolmates, co-workers, neighbours, fellow members of clubs, professional advisers, consultants, etc.) with whom one discusses important matters or spends time.

Mapping relationships

Two people can have an infinite set of possible relations. They may share attitudes, ideas, or gender. They may be friends, relatives, or co-workers. They may be sexual partners or simply play in the same football team. Furthermore, two or more of these connections may simultaneously occur between the same couple of people. Some of them are cooperative relations, while others convey open hostility, with an entire spectrum possible in between. Finally, some can be perceived only by one side and ignored by the other (e.g. the fans of a rock star feel linked to him, while the star may simply ignore them). Sociology has classified a broad range of possible links between individuals (Table 1). The tendency to have several kinds of relationships in social networks is called *multiplexity*. But this phenomenon appears in many other networks: for example, two species can be connected by different strategies of predation, two computers by different cables or wireless connections, etc.

We can modify a basic graph to take into account this multiplexity, e.g. by attatching specific tags to edges. For instance, we can take into account whether a connection is positive or negative. Species are linked by predation (negative), but also by mutualism (e.g. the positive relation established between flowering plants and pollinators). People can be enemies (negative) or friends (positive). A web page can link to another web page to criticize its content (negative) or to advertise it (positive). Adding this simple binary feature complicates things a lot. Imagine a group of three people, Alice, Bob, and Carol. When positive

Table 1. A classification of ties in social networks.

Similarities			Social relations				Interactions	Flows
Location	Membership	Attribute	Kinship	Other role	Affective	Cognitive		
e.g.,	e.g.,	e.g.,	e.g.,	e.g.,	e.g.,	e.g.,	e.g.,	e.g.,
Same spatial and temporal space	Same clubs	Same gender	Mother of	Friend of	Likes	Knows	Sex with	Information
	Same events	Same attitude	Sibling of	Boss of	Hates	Knows about	Talked to	Beliefs
				Student of		Sees as happy	Advice to	Personnel
				Competitor of			Helped	Resources
							Harmed	

relations connect all of them, everything is fine. Alternatively, Alice and Bob may be linked by friendship but both have hostile relations towards Carol. Things get complicated when the situation is reversed: Alice has positive relations with both Bob and Carol, but the two of them hate each other. And things get really problematic when everybody hates everybody else. According to sociology, the first and the second situation are *structurally balanced*, while the third and the fourth are not. In 2006, mathematician Tibor Antal and physicists Paul Krapivsky and Sidney Redner applied this concept to the shifting diplomatic alliances between six European countries before the First World War. They showed that their alliance gradually evolved into a structurally balanced situation, where either strong alliances were established or clear common enemies were identified. The six countries became divided into two groups (on one side, Britain, France, and Russia; on the other, Austria-Hungary, Germany, and Italy), each of them allied to all the countries in its group and enemy to all the countries in the other group. Soon after this situation had come about, the war broke out. This example shows that structural balance is not necessarily something desirable.

Graph theory allows us to encode in edges more complicated relationships, as when connections are not reciprocal. Wolves predate sheep, blogs link to large newspapers, and some people fell in love with others; the reverse is seldom true. In this case, the connections in the graph are a sort of one-way street where we can travel in one direction but not backwards. If a direction is attached to the edges, the resulting structure is a *directed graph* where links are indicated by an arrow. In these networks we have both *in-degree* and *out-degree*, measuring the number of inbound and outbound links of a node, respectively.

The relations considered so far are binary: that is, they acquire only two values. Such *dichotomous* connections either exist or do not: for example, being married to somebody or being employed by somebody. Statistically, however, these are exceptions: in most cases, relations display a broad variation of intensity. Predation is counted in the number of prey eaten, web pages can be connected by a sporadic link or by a large number of connections, and love can range from slight attraction to furious passion. Such further features correspond to the *weights* we can add to the links. *Weighted networks* may arise, for example, as a result of different frequencies of interaction between individuals or entities.

Other modifications of the basic graph structure are possible, and the techniques to handle these objects are very interesting. For example, a large part of social network research is devoted to working out how different kinds of ties affect each other. However, the strength of the network approach is that, in some cases, it is justifiable and effective to ignore all or most of the specific details: directed networks become undirected, weights are removed, multiple links are collapsed in a single edge, etc. Results show that this radical simplification can still capture a remarkable amount of information.

Chapter 3
A world of networks

Networkomics

During the eighties and nineties of the last century everything was 'genetic' in some way. Newspapers published stories about 'the gene for homosexuality', 'the gene for obesity', 'the gene for violence', or the 'gene for alcoholism'. This attitude responded to the expectation that the secret of human complexity was hidden in the genome. The DNA—the deoxyribonucleic acid molecule packed in the nucleus of the cell, that contains the genes—was dubbed the 'software of life', the program responsible for every single feature of a living being, the code whose dysfunction caused all the diseases. This vision set off a rush to sequence the genome, culminating in the publication of its map in February 2001. Results were quite surprising. Human beings do not have many more genes than a nematode worm, and fewer than some species of rice. It is reasonable that the human genome is almost identical to that of great apes, but the problem is that it is also rather similar to that of mice. The software metaphor did not stand in the face of this evidence: the DNA sequence alone does not explain the observed differences between species, let alone all the features and diseases of a single individual. In fact, there is a long series of steps from the genes to the macroscopic features of a living being. Variations in this path determine different outcomes.

The first layer of complexity above the gene level is given by *gene regulation*. Genes contained in the DNA are transcribed and translated to produce proteins. Proteins play a central role in almost every aspect of life: muscle movement, blood circulation, acting as enzymes, binding to hormones, etc. Moreover, proteins interact with each other: the production of a protein can be facilitated or hindered by the presence of other proteins in the cell. The delicate balance of these reciprocal influences is crucial for life. For example, the mutations of a single protein, the p53, are implied in a large number of different cancers. These interwoven patterns of activation and inhibition yield the *gene regulatory network*. In this net, nodes are genes and links are chains of reactions that connect the expression of a gene with that of others.

Protein interact, in all the forms in which they can happen, represent a second layer of complexity. For example, several proteins can bind together. These macromolecules behave as molecular machines, performing functions in the machinery of the cell. To do so, they must have the correct geometrical shape to fit with each other. When a protein is folded in the wrong way, several problems can arise. For example, the proteins responsible for the 'mad cow syndrome' (Creutzfeldt-Jacob disease) in humans, i.e. the *prions*, are supposed to be nothing else than misfolded proteins. All the possible physical connections between proteins can be represented as a network. In the *protein interaction network* the vertices are proteins and an edge is drawn between them if they physically interact in the cell.

Proteins are not enough for making a cell work. The cell interchanges matter, energy, and information with the environment, through many different molecules, involved in millions of reactions. Hunger, satiety, coldness, and in general all the states experienced by the organism, depend on this set of reactions, called *metabolism*. The chains of reactions that convert one molecule into another, passing through a series of intermediates steps, are called *metabolic pathways*. However, reactions in cells rarely follow the pattern of an ordered sequence. For example, the final molecule often interacts with the initial one in order to stop the reaction. This feedback process closes a loop in the chain of reactions. The ensemble of all such paths yields an intricate *metabolic network*.

An organism is therefore the outcome of several layered networks and not only the deterministic result of the simple sequence of genes. Genomics has been joined by *epigenomics*, *transcriptomics*, *proteomics*, *metabolomics*, etc., the disciplines that study these layers, in what is commonly called the *omics revolution*. Networks are at the heart of this revolution.

Thinking webs

The idea that the 'soul' could be embodied in an organ sounded a weird supposition by the 18th century. But physicians were aware that a stroke or other brain injuries could compromise crucial cognitive functions: the link between mind and brain was then starting to become evident. At that time, the anatomist Franz Joseph Gall dared to propose that all mental functions must arise from the brain. He identified 27 'organs' within the brain, each one responsible for colour, sound, memory, speech, as well as friendship, benevolence, pride, etc. The idea sounded so heretical that Gall had to flee Vienna and find shelter in revolutionary France.

Later on, several physiologists tried to verify Gall's theory, for example by removing slices from pigeons' brains. However, they could not find any evidence of the organs that Gall postulated. For this reason, they arrived at the conclusion that the brain was a homogeneous, undifferentiated unity that generated thought: 'the brain secretes thought as the liver secretes bile', as one of them put it. This conception dominated until the studies of Paul Broca in the 1860s. In autopsies of patients with expressive aphasia, Broca always found some damage to the frontal lobes of the left side of the brain.

'We speak with the left hemisphere,' he declared, after having identified what is now called *Broca's area*. Since then, neurologists have found various centres responsible for different activities, but they have also found that they rarely work in isolation: the integration of different areas of the brain is crucial to its functioning.

Networks provide a bridge between the paradigms of a brain divided into specialized areas versus the model of the brain as a whole (not dissimilar to what happens in social sciences, where networks allow us to describe society at a level between individuals and communities). The brain is full of networks where various web-like structures provide the integration between specialized areas. In the cerebellum, neurons form modules that are repeated again and again: the interaction between modules is restricted to neighbours, similarly to what happens in a lattice. In other areas of the brain, we find random connections, with a more or less equal probability of connecting local, intermediate, or distant neurons. Finally, the neocortex—the region involved in many of the higher functions of mammals—combines local structures with more random, long-range connections. Some scientists think that these wiring schemes may be responsible for subjective awareness: the emerging conscience may be the result of a sufficiently complex network structure.

Pinning down the actual structure of these neuronal networks is extremely hard, due to the enormous quantity of cells and the difficulty of probing them. Only for very simple organisms such as the nematode worm, *Caenorabditis elegans*, we have a detailed map. This one-millimetre long, transparent creature, with a three-week lifespan, has only about 300 neurons but it is a superstar in molecular biology. *C. elegans* is a *model organism*, an animal that is especially suitable for experiments, because scientists know its features well, and some aspects of it are comparable to those of the human organism. This translucent worm is often the first benchmark for the trial of new medicines and treatments.

Drawing a similar neuronal network for the human brain is impossible at the moment. However, another strategy can be applied. When humans perform an action, even one as simple as blinking, a storm of electrical signals from the neurons breaks out in several areas of the brain. These regions can be identified through techniques such as *functional magnetic resonance.* Through this technique, scientists have discovered that different areas emit correlated signals. That is, they show a special synchronization that suggests that they may influence each other. These areas can be taken as nodes and an edge is drawn between two of them if there is a sufficient level of correlation. Also at this level, the brain appears as a set of connected elements. Each action of a person lights up a network of connected areas in their brain.

The blood vessels of Gaia

In 1999, San Francisco Bay experienced massive algal blooms. Normally, such blooms are the result of an intensive agricultural use of land: when we drain fertilizers such as nitrogen and phosphorus into the sea, they become nutrients for algae. However, this was not the case in this instance, since a number of policies and controls had decreased the level of nutrient pollution entering the bay from its various rivers. Compiling data from three decades of observations, ecologists in California concluded that the blooms had a much more complicated explanation. In 1997 and 1998, one of the strongest *El Niño* events was recorded, followed by an equally strong *La Niña* in 1999. These phenomena induced changes in the California current system. Deep, cold, and nutrient-laden waters emerged along the coast. Such nutrients attracted ocean dwellers—flatfish and crustacean—into the bay. These animals are predators of bivalves of the bay that, in turn, act as an obstacle to the spreading of algae. The collapse in the bivalve population due to the increase in its predator was the immediate cause of the algal blooms. The conditions that triggered this domino effect may be due to normal climatic fluctuations. Nevertheless, its consequences are a warning: climate change, and especially the increased frequency of extreme events, can have rather unexpected effects on ecosystems.

The central structure behind the San Francisco Bay algal bloom is a *food chain*—that is, a series of species in connection: flatfish and crustacean prey on bivalves, and bivalves consume algae. Through food chains, living organisms extract from each other the energy and matter they need to survive (this is not the only possible interaction between species in an ecosystem: organisms can also establish mutually beneficial interactions, such as those between flowering plants and their insect pollinators). Every food chain starts with *basal* species, such as plants and bacteria. These do not prey on any other species and take resources directly from the environment by transforming light, minerals, and water. These resources are transferred along the food chain by successive predations. *Intermediate* species are organisms that are both predators and prey. And *top* species (at the end of the chain) are those that are not predated by anything. Food chains help us understand why fisheries collapse, as the Peruvian anchovy fishery in the seventies. After periods of massive indiscriminate fishing, the result is a dramatic reduction of predators such as cod or tuna. After this phase, fishing tends to move towards more basal species, such as anchovies. But these rapidly collapse as well. The reason is that, when large predators are removed, they are replaced by other predators downstream in the foodweb. These are often non-edible fish: without population control, they deplete the other edible basal species.

The actual picture of an ecosystem is even more complicated: typically, food chains are not isolated, but interwoven in intricate patterns, where a species belongs to several chains at the same time. For example, a specialized species may predate on only one prey (or in some cases on only a few). If the prey becomes extinct, the population of the specialized species collapses, giving rise to a set of *co-extinctions*. An even more complicated case is where an omnivore species predates a certain herbivore, and both eat a certain plant. A decrease in the omnivore's population does not imply that the plant thrives, because the herbivore would benefit from the decrease and consume even more plants.

As more species are taken into account, the population dynamics can become more and more complicated. This is why a more appropriate description than 'foodchains' for ecosystems is the term *foodwebs* (Figure 5). These are networks in which nodes are species and links represent relations of predation. Links are usually directed (big fishes eat smaller ones, not the other way round). These networks provide the interchange of food, energy, and matter between species, and thus constitute the circulatory system of the biosphere. They are the blood vessels of the Earth.

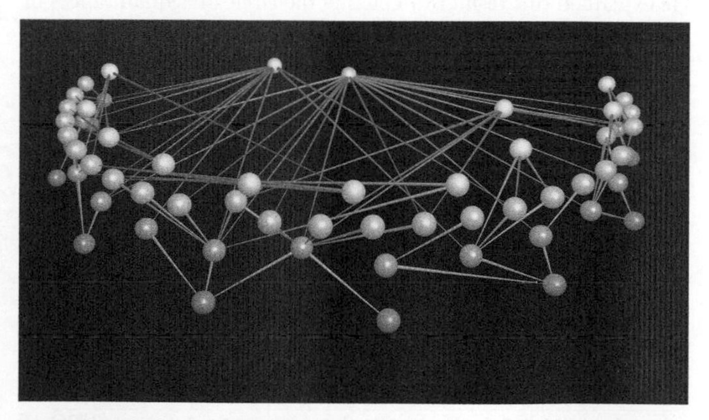

5. A United Kingdom Grassland foodweb: nodes represent species of grassland plots in England and Wales, and links are drawn from predators (thicker end) to preys (thinner end)

Homo 'retiarius' (Net man)

Word of mouth is a common way to obtain information about white-collar job openings. So if we are looking for this kind of job, it is a good idea to spread the word between friends and relatives. Less obviously, it may be even better to inform distant acquaintances and people we do not see often. This is what Mark Granovetter suggested in 1973. This sociologist interviewed a sample of professionals in a Boston suburb who had recently relied on personal contacts to obtain their jobs. He asked them how often they saw the person before obtaining the job. The majority reported 'occasionally' and a significant fraction answered 'rarely'. Job offers are more likely to come from old college friends, past workmates, and previous employers, than from close friends. Chance or mutual friends were the channels by which these connections were rediscovered. Granovetter described this phenomenon as *the strength of weak ties*.

He explained this result by depicting the circle of acquaintances of a hypothetical individual called Ego. Ego lives every day with his family and some close friends. Probably all these persons are also in close contact with each other. As a result, information travels fast in the group. So Ego is likely aware of all the news available in the group. On the contrary, weak ties connect him to faraway people. These individuals are not bounded by Ego's social surroundings. Therefore they open a whole set of new groups to him, each of them encapsulating information otherwise inaccessible. Missing the opportunity of weak ties causes difficulties in organizations, companies, or institutions. Information and skills become trapped in one group, without reaching those who need them. So these things have to be reinvented or paid for from outside consultants. A former CEO of HP is reported to have lamented: 'If HP only knew what HP knows!' Granovetter's intuition was later developed into the theory of *social capital*. This idea implies that the contacts of one person (and the contacts of these contacts) enable him or her to access resources that ultimately provide such things as better jobs and faster promotions. More generally, the position of an individual in his or her social network is crucial to determine future opportunities, constraints, outcomes, etc.

Measuring acquaintance is not easy, since it is such a subjective issue. Usually, maps such as the flowchart of a company are not very useful, because they do not correspond to the actual relations between workers. For this reason they are useless in helping us understand the channels (and possible bottlenecks) of information within the company. Scientists have devised a large set of alternative strategies to draw social networks, ranging from questionnaires to *snowball sampling*, a system in which an interviewed subject suggests somebody in their circle for the next interview. These strategies have enabled the collection of data as sensitive as the map of sexual intercourse between groups of different individuals (from high-school students in the US Midwest to people living in a village of Burkina Faso): the knowledge of these networks allows for better understanding of the spread of sexually transmitted diseases.

Another kind of relation that is relatively easy to pin down is professional collaboration. Such networks exist in several fields, ranging from Hollywood (two actors become connected if they play in the same movie) to science (two scientists become linked if they write a paper together). Collaborations can be found in more exotic environments such as politics (US senators have been connected on the basis of the co-sponsorship of laws) or terrorism (activists are connected on the basis of intelligence reports and legal documents).

Information technology provides a new and powerful way to measure interaction between people. Frequent phone calls and emails between two individuals, or friendship in virtual social networks like Facebook or LinkedIn, indicate a stable relationship and therefore an edge. More and more companies exploit the social networks of their customers to find such information. For example, telephone companies are reported to target 'influential' individuals with offers and other strategies: these are the customers who, when they change company, trigger similar changes in their close connections.

Webs of words, webs of ideas

'What shall King Henry be a pupil still / Under the surly Gloucester's governance?' says Queen Margaret in Part II of Shakespeare's King Henry VI trilogy (Act I, Scene III). The queen complains about the influence of the duke of Gloucester on her husband, the king. What does she mean by 'pupil'? A thesaurus suggests that *pupil* can be rephrased as scholar, acolyte, adherent, convert, disciple, epigone, liege man, partisan, votarist, or votary. This list provides a full spectrum of words to denote one person subjected to another. We can enlarge the spectrum by exploring all the words related to 'pupil' that is, its *semantic area*. These include faithful, loyalist, advocate, backer, supporter, satellite, yes-man . . . What role does the queen desire for her husband? Antonyms of 'pupil' suggest non-student, coryphaeus, leader, apostate, defector, renegade, traitor, and turncoat. Naturally, she is asking King Henry to rebel against the authority of the duke of Gloucester.

This is a simple example of how words link to each other. Indeed, a rigorous analysis should take into account the historical differences in the use of a word, the specific occurrence of a term in Shakespeare's work, the context in which it is used in the script, and many other aspects. In any case, it is fair to say that we can better understand the meaning of a word if we take into account its 'neighbours' in language. *Synonimity*, *antonymity*, and *semantic connection* are just a few of the possible relations. Others are *meronymity* and *hypernimity* ('beer' is a meronym of 'drink' and the latter is a hypernym of the former). 'Pupil' provides a crystal-clear example of another kind of connection. This term has two completely different meanings: it indicates both a student and a part of the eye. It is a case of *polysemy*. Of course, the context of Shakespeare's dramas immediately determines the correct meaning. In general, context provides the exact meaning of words: the *co-occurrence* of words in sentences defines their meaning. Such a co-occurrence provides yet another relation between words. For example, the words 'king' and 'Henry' are much more likely to appear together in English sentences than 'king' and relativity, for example.

We can now create our own maps of language. We use words as vertices and the edges connect synonyms, antonyms, and polysemic words (these relations can be drawn from a thesasurus or a dictionary, while patterns of co-occurrence can be drawn from large language databases, such as the *British National Corpus*). Semantic connections are more difficult to pin down: their study forms a complete area of linguistics. Some languages have special dictionaries that associate one word with a set of related ones. An alternative approach is experimental *word association*. A word is provided to a sample of people, asking them to say the first word that comes into their mind after hearing it. The resulting words are then used to repeat the association experiment. Proceeding in this way, step by step we build our web of associations. Different instances of word networks display different results. These depend on the language, on the kind of text, on the education of the author of the text, or may be related to linguistic dysfunctions.

Word networks contain a lot of information, but usually they are not particularly useful in studying the actual content of the texts and the relations between the ideas expressed in different texts. This is a crucial issue for web queries, for example. Usually, complicated algorithms have to be implemented to perform this task. However, in some bodies of texts, a very precise network can be drawn. This is the case with scientific literature. Producing knowledge is never a solitary endeavour. A scientist is always 'a dwarf standing on the shoulders of a giant', as philosopher Bernard of Chartres is said to have pointed out for the first time in the 12th century. Scientists' work almost always builds on previous results. Researchers recognize this by citing several older publications at the end of their papers. Citations provide recognition of relevant results of the past, give credibility to new results and make reference to facts, technologies, and experiments that are accepted as valid, or criticized, within a work. Publications have been extensively standardized in recent years: articles are mostly in English, control methods have been homogenized (mainly through *peer review*), measures of impact have been devised, etc. At the same time, large electronic databases of publications have been established, with thousands of new items being added every day: articles, books, patents, projects, etc. All this produces a large network of publications: two items are connected if one of them cites the other. We can also identify authorship from these databases and create networks of collaborations between scientists. These systems are increasingly used to map and visualize the development of knowledge and the most active areas of science.

Money by wire

In 2008 a number of large financial institutions in the United States suddenly went bankrupt. In a few months, the majority of the developed world was involved in one of the largest financial crises ever seen. Much has been written on the causes of this crisis. What is certain is that it showed that economies are very tightly interconnected at a global level.

Classical economic theories represented economic actors as independent, completely rational agents, focused on maximizing their income. However, facts show that individuals, companies, institutions, and countries are not independent: everyone influences each other in many ways. Their behaviour, far from being completely rational, is strongly dependent on subjectivity, emotions, and reciprocal influence.

Lending money is one way in which companies and institutions can become tightly interconnected. An interesting case is the money exchanged on a daily basis between private banks so that it can be available to meet the possible requests of bank clients (becoming therefore *more liquid*). If customer requests should exceed the liquidity reserve of a bank, that bank can ask other banks to lend money. Central banks worldwide require other banks to place a part of their deposits and debts with them, to create a buffer against liquidity shortages. In this sense, central banks ensure the stability of the banking system, thus avoiding liquidity shocks. The freezing of the interbank lending network was one of the first signals of the 2008 financial crisis.

An even stronger relation than lending money is given by *shareholding*, i.e. the direct participation of a company in another company's capital. This means that the first company holds a part of the second one, and can exert influence on its main decisions. Shareholding is converted into control when a company holds the majority of stock, or when it is able to determine the vote of the majority of the board. In this case, legally independent companies are converted into a business group. Often, these groups display a pyramidal structure, where a *holding* company is at the top, and operating companies are at the bottom of the control hierarchy.

The existence of business groups is explicit and legally regulated in most countries; but softer and less regulated forms of influence can exist. The most common of these happens within boards. Managers often sit on many boards at the same time. Obviously, they act as channels of information, alliances, or interests between boards. Their simultaneous presence in different boards establishes an *interlock* between their companies. If the companies are explicit competitors, this situation is clearly incompatible with a free market. A shared director will either favour one of the companies against the other or establish a cartel between them (which is generally ruled out by law). In general, such a director will find it very difficult to operate in the interest of all the investors in the different companies.

Further evidence of interconnectedness between companies is given by *stock price correlations*. Finance practitioners know that the stocks of companies operating in the same sector (e.g. mining, transport, services, food, etc.) change their prices in a somehow similar or 'synchronized' fashion. For example, stock prices of all the companies within the electronic sector (or any other) tend to decrease or increase at the same time. Financial analysts are interested in knowing how much of the change in a stock price is influenced by the change of another stock (in short, they want to know the *correlation* between stock prices). If this connection is strong enough, it is likely that the two companies are somehow connected. Lending money, shareholdings, shared directors, or stock price correlations are the main criteria by which a network between companies can be built: edges are drawn when one of these situations occurs.

The interconnectedness goes far beyond companies within one specific market. As the financial crisis has shown, events rapidly spread from national markets into the global scenario. One obvious channel by which this can happen is the import/export trade relationships among countries. This *world trade web* is a network in which nodes are countries and their trade relations define the edges. Like the cell, economies depend on these multiple layers of networks.

Critical infrastructures

On the night of 28 September 2003, lights went out across the whole of Italy, with the single exception of the island of Sardinia. It took several hours, in some places even days, to reinstate normal supply. Investigation showed that the blackout was triggered by a tree flashover close to a high-tension line between Italy and Switzerland. The resulting shortage of electric supply caused a sharp increase in demand on the remaining lines. As a result those lines collapsed, generating a ripple effect through the entire system.

Large-scale power outages reveal the connectedness of power grids. These systems deliver electricity across large distances from central points to cities and industrial areas. Carefully planned in the beginning, they grow more and more intricate over time. Nowadays, generators, transformers, and substations, connected by high-voltage transmission lines, constitute a network that spans several regions and often several countries (as the 2003 example shows). It is clear that such networks require careful maintenance to prevent criticalities.

Similar instabilities appear in a variety of other infrastructures. Communication systems, such as the telephone network, are one example. But probably the most sensitive is the transportation network: streets, highways, and railways connecting cities, the web of boats transporting fuel and other goods, and above all the airport web. Planes transport billions of passengers and tonnes of goods every year. A minimal malfunction in such an infrastructure has major consequences: Eurocontrol (the European organization for the safety of air navigation) has estimated that delays in flights cost European countries up to €200 billion in 1999 alone. In a globalized world, transportation networks are similar to circulatory systems in living beings.

A net as large as the world

In October 1969, a message travelled for the first time from one computer to another, through a telephone line. Two university labs in California were at the ends of the line. After a few letters, the message broke down, but the connection was established: Arpanet, the grandfather of the Internet, was born. The idea of a network of computers was around during the previous decade. At the end of the fifties, ARPA (the US Advanced Research Project Agency) asked engineer Paul Baran to design a communication structure able to resist an attack. In particular, the whole system had to continue working even under an attack destroying part of it. Baran duly designed a distributed system with such characteristics—but a change in strategy locked his pioneering studies into a drawer. However, in the sixties, some universities asked ARPA to finance a similar project for different purposes. The academic institutions were keen to interconnect their computers in order to aggregate their computational power.

The 1969 Arpanet connected UCLA (University of California Los Angeles) and the SRI (Stanford Research Institute). Two years later, the number of nodes was over forty, including some companies and other universities. This structure was so successful that in the seventies similar networks appeared in other parts of the world, created by particle physicists, astronomers, companies: Hepnet, Span, Telenet, etc. If at the beginning the problem was to connect the computers, this moved on how to connect networks. *Internetworking* became the motto of many computer scientists. At the end of the seventies, engineer Robert Kahn and mathematician Vinton Cerf developed the TCP/IP: whatever the internal structure of networks, this software allows them to talk to each other. This code was put in the public domain and based on the concept of *open architecture*. Finally, in the eighties, the *TCP/IP transition* was fully achieved, bringing to creation the Internet, the 'network of networks'.

Such a structure is probably the human artefact that best embodies the idea of networks. A computer connected to the Internet becomes one of many *hosts*. If we want to deliver an email to a specific place, we do not need to be directly connected with that destination. From origin to our target, information travels along *routers*, devices responsible for transmitting packets of data. A large series of connections keeps the structure linked: optical fibres, telephone lines, satellite connections, etc. Since nobody plans where hosts and connections are added, the overall structure of the Internet is not recorded. Actually, mapping at host level is practically impossible. We can have a rough representation of it only at router level. In this case, the nodes of these networks are the routers, and the edges are their connections. We can coarse-grain the structure even more, grouping routers into *autonomous systems*. These groups are autonomously administrated domains that usually correspond to Internet Service Providers (ISPs) and other organizations.

The great success of the Internet is due to the exceptional experience it provides. Watching television is a one-direction, one-medium, passive experience. The Internet is not so. People can navigate an infinite series of documents, use different media, exchange information, and talk to each other. Unlike traditional communication technologies such as telephone, radio, or television, the Internet does not have a specific purpose. Rather, it is a mutant artefact able to host infinite applications. Actually, the Internet is only a physical infrastructure that supports services. One of the most successful of these is the World Wide Web (WWW). This is an enormous set of electronic *documents* recorded in the devices that make up the Internet and connected by *hyperlinks* that allow navigation between them. The pattern is somewhat similar to the body of scientific literature, made of articles, books, patents, etc. connected by citations.

The idea of the WWW was born at CERN (European Organization for Nuclear Research). Physicist Tim Berners-Lee (later joined by computer scientist Robert Cailliau) put forward the proposal for it in 1989. Berners-Lee designed a system that allowed scientists to access, through their own computers, the enormous amount of data produced by particle physics experiments. The software to make this system work was not patented but rather released in the public domain. This decision—as in the case of TCP/IP—proved to be very significant. Right from the beginning, thousands (and then even greater numbers) of users tried it, improved it, and created web pages and services. In just a few years, the Web became World Wide. Its magnitude is unknown, since none of the search engines that explore the web (such as Google or Yahoo!) is able to archive all web pages. After all, this would make little sense: several websites are capable of producing new pages upon request. A 2005 estimate put the content of the whole WWW (for static pages) as equivalent to 200 *terabytes* of information. At the time, this was about ten times the size of the US Congress Library. Undoubtedly today's figure would be greater by orders of magnitude, since the growth of the WWW is exponential.

Cyberspaces

On 11 September 2001, the infrastructures of New York City experienced a 'network catastrophe', parallel and related to the human tragedy that was taking place that same day. Soon after two hijacked airliners crashed into the World Trade Center complex, a surge in phone calls was registered. People were trying to communicate with friends, and rescue personnel with colleagues. Cell networks were quickly overloaded and people lined up at payphones in Manhattan. The attack damaged Verizon's central office, interrupting 200,000 lines. AT&T infrastructures, some of them housed in the basement of the World Trade Center were destroyed as well. When calls failed, many people turned to the Internet. But wireless service was also impacted. The impact on the economy extended well beyond the crash sites. It took six days before the New York Stock Exchange could return to work. Months passed before services were restored to near pre-disaster levels.

The terrorist attacks showed that there is almost no network that stands alone. Physical and virtual infrastructures are embedded in a common *cyberspace*, where energy, information, transport, communication, etc. are provided. Power grids support the Internet, that hosts the WWW, that in its turn enables email services, social networks, information websites, and file-sharing systems. Many activities, including flight control, bank programs, emergency systems, and commercial services also depend on the WWW. A collapse at one level of cyberspace can affect the other layers, often in a quite unpredictable way.

The interconnection of multiple networks is common in many other situations. For instance, the liquidity shock that affected the economy in 2008 rapidly spread to many other economic networks. Similarly, social networks show this feature in many aspects. One interesting example is that of real-world friendships as compared with virtual social network contacts: non-trivial feedbacks exist between the two networks. Cells are a miniature cyberspace: genetic regulatory networks, protein interaction networks, and metabolic networks are nested into one another; for this reason, some scientists have proposed fusing the concepts of genome, proteome, metabolome, etc. into the comprehensive idea of *interactome*. Finally, ecosystems can be seen as sets of interacting networks; for example, both antagonistic and mutualistic networks play a role in determining how species will thrive. In all these cases, networks provide useful maps to disentangle complex and interwoven systems.

Chapter 4
Connected and close

One world

On 4 November 2006, the disconnection of a single power line in north-west Germany triggered an avalanche of blackouts as far away as Portugal. One expects that such a small disconnection could have effects at the regional level, or at most within the national power grid. But electrical networks have become more and more integrated (currently, they form one large system at continental level) and then more fragile. The same has happened in other infrastructures. For example, almost all airports are interconnected: today it is possible to reach practically every destination, starting from any point of origin, with a limited number of stops on the way. The Internet is fully connected as well, since it has grown out of the integration of smaller networks.

Networks in nature and society may very well be fragmented in separated parts. In the cell, some groups of chemicals interact only with each other and with nothing else. In ecosystems, certain groups of species establish small foodwebs, without any connection to external species. In social systems, certain human groups may be totally separated from others. However, such disconnected groups, or *components*, are a strikingly small minority. In all networks, almost all the elements of the systems take part in one large connected structure, called a *giant connected component.*

For example, in one experiment based on mental associations between words, scientists found that 96 per cent of the terms ended up in one large group. One could find a path between any two expressions in this group, even such different terms as 'volcano' and 'stomach': the chain of associations drawn by the people participating in the experiment was 'volcano', 'Hawaii', 'relax', 'comfort', 'pain', and 'stomach'. In general, the giant connected component includes not less than 90 to 95 per cent of the system in almost all networks. A few interesting consequences of this fact can be listed. In sexual interaction networks, past and current intercourse can connect us with individuals we would never imagine or desire to be related to. In the network of collaboration between scientists, a large cooperation structure appears, omitting only a few lonely players. In company boards, interlocking grants a kind of connectedness that encompasses the vast majority of companies. Finally, in foodwebs, pollutants introduced in one species at a given location are brought by foodchains to apparently unrelated species as far away as on the opposite side of the planet.

Living in a big, connected world does not always imply that every node can be reached from every other. As with ordinary roads, where it is crucial to know if they are one-way or not, we must know if edges are directed. In a directed network, the existence of a path from one node to another does not guarantee that the journey can be made in the opposite direction. Wolves eat sheep, and sheep eat grass, but grass does not eat sheep, nor do sheep eat wolves. This restriction creates a complicated architecture within the giant connected component (Figure 6). For example, according to an estimate made in 1999, more than 90 per cent of the WWW is composed of pages connected to each other, if the direction of edges is ignored. However, if we take direction into account, the proportion of nodes mutually reachable is only 24 per cent, the

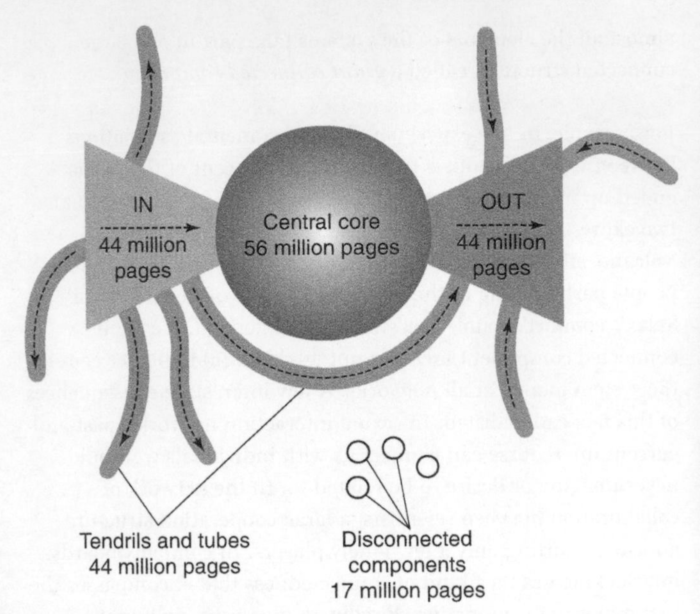

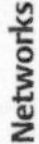

6. Networks with directed links, like the WWW, display a bow-tie structure: a central giant strongly connected component, an in-component, and an out-component, as well as minor structures (tubes, tendrils, and a few disconnected components). Data in the picture are from 1999.

giant strongly connected component. The rest is divided between an *in-component* and the *out-component*: the first is made up of pages with paths pointing to the giant strongly connected component, and the second of nodes that receive links coming out from it (the picture is completed by minor structures called *tubes* and *tendrils*). This characteristic structure gives the giant connected component of the WWW a characteristic 'bow-tie' shape. This complicated structure is not specific to the WWW, but appears, with different components, in all directed networks.

The existence of the giant connected component—either with a bow-tie structure or not—is a remarkable feature. For example, the giant strongly connected component of the WWW, although being less than one-third of the system, still encompassed 56 million pages in 1999. An obvious explanation of this feature would arise if networks were very *dense*—that is, if they had many redundant links, sufficient to connect every node with almost every other. Yet usually this is not the case. On the contrary, most networks are *sparse*, i.e. they tend to be quite frugal in connections. Take, for example, the airport network: the personal experience of every frequent traveller shows that direct flights are not that common, and intermediate stops are necessary to reach several destinations; thousands of airports are active, but each city is connected to less than 20 other cities, on average. The same happens in most networks. A measure of this is given by the mean number of connection of their nodes, that is, their *average degree*. We create many web pages every day, but the average number of links is about ten. Hundreds of thousands of routers connect the Internet, but each of them connects to fewer than three others, on average. Finally, in a sample of more than 50,000 physicists, the average number of collaborators was found to be about nine. In most real-world networks, there may be elements with many connections (and indeed, there are), but in general the graphs are not dense: on the contrary, they are said to be *sparse*.

This puzzling contradiction—a sparse network can still be very well connected—had already attracted the attention of the Hungarian mathematicians we encountered in Chapter 2, Paul Erdős and Alfréd Rényi. They tackled it by producing different realizations of their random graph. In each of them, they changed the density of edges. They started with a very low density: less than one edge per node. It is natural to expect that, as the density increases, more and more nodes will be connected to each other. But what Erdős and Rényi found instead was a quite abrupt transition: several disconnected components coalesced suddenly into a large one, encompassing almost all the nodes. The sudden change happened at one specific critical density: when the average number of links per node (i.e. the average degree) was greater than one, then the giant connected component suddenly appeared. This result implies that networks display a very special kind of economy, intrinsic to their disordered structure: a small number of edges, even randomly distributed between nodes, is enough to generate a large structure that absorbs almost all the elements.

So close

In early 1994, three students at Albright College—Craig Fass, Brian Turtle, and Mike Ginelli—were watching television during a heavy snowstorm. According to their own reports, they noticed that Kevin Bacon, whose next movie was announced on television, was present in many different movies. So they began to speculate about the large number of actors that had starred with him in those movies. The idea that Bacon was a kind of 'center of the entertainment universe' started to spread, became famous, and a web page based on it even appeared, the *Oracle of Kevin Bacon*: this search engine provides the relation between Bacon and any actor one may search for. Remarkably, if one writes the name of a Spanish actor from old commercial movies, such as Paco Martínez Soria, the oracle finds a quite close relation: Martínez acted in *Veraneo en España* with Luís Induni; the latter acted in *Il Bianco, il Giallo e il Nero* with Eli Wallach; and Wallach acted in *Mystic River* with Kevin Bacon. Such short chains of collaborations with Bacon are found for almost any actor one may think about.

This surprising feature brings to mind a game played by scientists. Paul Erdős, the expert of random graphs, was one of the 20th century's leading mathematicians. Scientists assign themselves, as a badge of honour, a measure of their collaboration with him: those that have co-authored a paper with him are said to have an Erdős number 1; co-authors of his co-authors have Erdős number 2; co-authors of co-authors of Erdős' co-authors have Erdős number 3; and so on. However, only the lowest Erdős numbers are a real source of pride: more than 500 scientists were direct co-authors of Erdős'; and a few thousands have collaborated with this nucleus of co-authors. Finally, tens of thousands people have Erdős number 3 (G. C. is one such; M. C. has Erdős number 4), so it is not exactly something of special merit. Almost no scientist in any field has an Erdős number greater than 13.

Contrary to what one may believe, these notable results are nothing specific to Kevin Bacon or Paul Erdős. The former is not the centre of the entertainment world; nor is the second the hub of mathematics. If the same calculations are repeated with any other actor or scientist, similar results are found: very short chains connect apparently remote individuals. This fact gives an interesting insight into a quite common *cocktail party experience*: you are talking with a stranger and suddenly discover that he or she is your wife's schoolmate, or your brother's tennis partner, or your friend's neighbour. This discovery is usually hailed as a surprise ('It's a small world!'), but it may not be so unusual. Social systems seem to be very tightly connected: in a large enough group of strangers, it is not unlikely to find pairs of people with quite short chains of relations connecting them.

Six degrees of separation

In 1967, American psychologist Stanley Milgram undertook a set of memorable experiments. With the collaboration of Jeffrey Travers, he sent tens of letters to citizens in the Midwest (Kansas and Nebraska), chosen at random. In the message, he asked them to forward the letter to a person in Massachusetts (either, the wife of a divinity student at Cambridge or a stockbroker in Boston), the address of whom he did not provide. In case they did not know the recipient, he suggested they forward the letter to somebody they knew, that might for some reason be 'close' to the recipients. For every forwarded letter, another one had to be sent to Milgram himself, so that he could follow the path of the messages. In a country of hundreds of millions, finding somebody by essentially a word-of-mouth procedure seems impossible. However, after a few days, the recipients started to receive the first letters. These had passed through only one intermediary. A few weeks later, when the experiment was declared finished, about a third of the letters had arrived at their destination: none had been sent more than ten times and, on average, the number of mailings was six.

The experiment sparked enthusiasm in the scientific community. In the 1950s, Manfred Kochen, a mathematician, and Ithiel de Sola Pool, a political scientist, had speculated that humans may be much 'closer' to each other than expected. They asked: If two persons are selected at random from a population, what are the chances that they would know each other? More generally, how long is the chain of acquaintanceship that is required to link them? In a highly circulated paper, which they eventually published in 1978, they proposed a mathematical model suggesting that, in a population such as that of the United States, an unexpectedly large fraction of pairs could be linked by chains with just a few intermediaries. Milgram's experiment was an empirical test of their intuition. This finding had a strong impact even beyond the academic world. The expression *six degrees of separation*, referring to the number found in Milgram's experiment, became the popular formulation of his results. In 1990, playwright John Guare used it as the title of a comedy in which a charismatic character dodges people, saying that he is the son of actor Sidney Poitier. Here is how Guare expressed Milgram's results:

> I read somewhere that everybody on this planet is separated by only six other people. Six degrees of separation. Between us and everybody else on this planet. The President of the United States. A gondolier in Venice. Fill in the names. [. . .] It's not just big names. It's anyone. A native in a rain forest. A Tierra del Fuegan. An Eskimo. I am bound to everyone on this planet by a trail of six people.

Small worlds

Iloveyou was one of the most contagious computer viruses ever. After appearing on May 2000, it hit tens of millions of computers all over the world, producing billions of euros of damage, mainly the costs of eradicating it. *Iloveyou* arrived in an email, disguised in an attachment that looked like a love letter. When the attachment was opened, the virus infected the computer, and forwarded itself to the email addresses contained in the address book of that computer. After just a few replications of this kind, the virus reached an enormous number of devices. As in the social networks described in the previous paragraph, one could say about the computer network where viruses spread: 'It's a small world!' A large number of computers are reached through just a few connections. Computers apparently far from each other turn out to be connected by short chains of links.

In reality, this *small-world property*, which is the main result of Milgram's experiment, is present in all networks. The Internet is composed of hundreds of thousands of routers, but just about ten 'jumps' are enough to bring an information packet from one of them to any other. Thousands of kilometers may stand between them, but that is not the distance that matters: what matters is the number of connections to be crossed, and this is always very small. That is why information travels across the planet at an extraordinary velocity. Another example is the WWW: it is composed of billions of pages, but scientists have found that about twenty mouse 'clicks' are enough, on average, to navigate between any two of them. There are fewer than three 'degrees of separation', on average, between any pair of neurons in the brain of a *C. elegans*. In the import–export network that interconnects the countries of the world, it is impossible to find two states separated by more than two links. The list of examples could go on with many other cases.

The small-world property consists of the fact that the average distance between any two nodes (measured as the shortest path that connects them) is very small. Given a node in a network (say, Paul Erdős in the co-authorship network), few nodes are very close to it (direct co-authors) and few are far from it (scientists with very high Erdős numbers): the majority are at the average—and very short—distance. This holds for all networks: starting from one specific node, almost all the nodes are at very few steps from it; the number of nodes within a certain distance increases exponentially fast with the distance. Another way of explaining the same phenomenon (the way scientists usually spell it out) is the following: even if we add many nodes to a network, the average distance will not increase much; one has to increase the size of a network by several orders of magnitude to notice that the paths to new nodes are (just a little) longer.

The small-world property is crucial to many network phenomena. The short synaptic distance in the neocortex may be crucial to its functioning: some studies suggest that neurodegenerative illnesses such as Alzheimer's imply a massive loss of the small-world property in the brain. Short distances in the sexual relations networks suggest that the concept of risk group in sexually transmitted diseases has to be interpreted carefully: virtually everybody is at a very short distance from somebody infected. This capability of networks to spread viral agents efficiently can also have constructive applications. One of the first examples of *viral marketing* strategy was the worldwide diffusion of the *Hotmail* email service, launched in 1996. People purchasing a free Hotmail address agreed to host in their messages a link that allowed recipients to open a free address in their turn. Hotmail had one of the fastest increases for a communication company, attacting tens of millions of users, partially due to this strategy that cleverly exploited the small-world nature of the email network.

Shortcuts

The small-world property is something intrinsic to networks. Even the completely random Erdős–Renyi graphs show this feature. By contrast, regular grids do not display it. If the Internet was a chessboard-like lattice, the average distance between two routers would be of the order of 1,000 jumps, and the Net would be much slower: no quick web browsing, no instant emails. If the network of scientific collaborations was a grid, Paul Erdős would have a moderate number of co-authors; these would have a larger, but still moderate, number of collaborators, etc.: the number of individuals within a certain distance would not grow exponentially but much more slowly. If the neurons network was a lattice, increasing the number of neurons (due to the physical growth of the brain, for example) would increase remarkably the average communication distance within the neocortex: paradoxically, growth would make people less smart (something young readers could agree with).

What makes a network different from a grid? What factor is responsible for the appearance of the small-world property in webs, while lattices lack it? In 1998, physicist Duncan Watts and mathematician Steven Strogatz tried to answer these questions. They started by considering a very simple, regular structure. It was a circle of nodes, and each of them was connected to its first- and second-nearest neighbours (Figure 7 left). This structure may represent a group of remote villages, each of them interchanging goods with the neighbouring villages, and occasionally with the neighbours of their neighbours. In such a regular structure, products can travel a long way from producers to consumers in faraway villages.

Watts and Strogatz then allowed for something very similar to opening a path between two distant villages. In practice, they cut

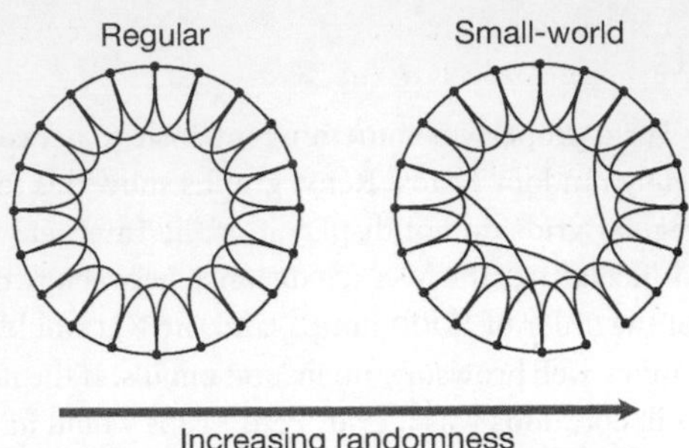

7. In the small-world network model a regular lattice is transformed into a network by introducing disorder, and correspondingly the distances between nodes drop: the small-world property arises

one of the links in the initial structure and rewired it with another node, chosen at random. Suddenly, the inhabitants of one village can interchange goods with a previously remote area, instead of their neighbours. Still, on the large scale, just a few villages are affected by this change, and several areas of the circle remain far from each other. One can see this by computing the average distance between nodes after the rewiring: it is just moderately reduced by the new shortcut. At this point the two scientists allowed for more 'paths' (Figure 7 right). After every rewiring, they computed the average distance: remarkably, they found that, after rewiring just a few links, the distance dropped abruptly. A small number of shortcuts is enough to bring all the elements of the system much closer to each other. The key ingredient that transforms a structure of connections into a small world is the presence of a little disorder. No real network is an ordered array of elements. On the contrary, there are always connections 'out of place'. It is precisely thanks to these connections that networks are small worlds.

These shortcuts are easy to identify in some nets. For example, in 1858, the first transatlantic telegraph cable connected Europe and

America. This wonder of thousands of kilometres and hundreds of tonnes is one of the marvels of the oceans visited by the submarine in Jules Verne's *Twenty Thousand Leagues Under the Sea*. Today, several transoceanic cables grant immediate spreading of information throughout the world. In word networks, polysemy is one main source of shortcuts. For example, the word 'pupil' connects the two semantic areas of teaching, on the one hand (pupil as student) and vision, on the other (pupil as the body organ). In social networks, Granovetter's concept of *weak ties*—links that connect unrelated groups—could be at least partially matched with Watts and Strogatz's shortcuts.

Shortcuts are responsible for the small-world property in many other situations. However, every now and then we can find another possible explanation. For example, the remarkably small distances in the world trade web are due to the fact that this is one of the few highly dense (non-sparse) networks: the average number of partners of a country is typically comparable with the total number of countries represented in this network, something that suggests that each one has interchanges with a large majority of the others. In foodwebs, other mechanisms favour the small-world property. Basal species get energy and matter from sun and environment, but their predators extract just 10 per cent, on average, of the resources contained in them, and the same happens at every successive step of predation: if foodchains were too long, top predators would not extract enough from them to survive.

Whatever its origin, the small-world property is a crucial feature to be taken into account when systems have the structure of a graph. The network approach provides a striking vision of such systems: in the first place, their elements are part of one big world, where almost every node has a path of connections with every other; in the second place, these paths are extremely short. This interwoven structure is essential to understanding a broad range of phenomena, from Aids to blackouts to information spreading.

Chapter 5
Superconnectors

Hubs

In his memorable experiments on the 'six degrees of separation', Stanley Milgram made an observation that was going to be fully understood only much later. In one of his experiments, the US psychologist asked random citizens of Nebraska to forward a letter to a stockbroker in Massachusetts. If they did not know the recipient, they should send the letter to somebody that they believed to be closer to him. Beyond the fact that a large part of the letters arrived in an average of just six steps, Milgram observed that a quarter of them were delivered to the recipient from the very same source: a clothing merchant and friend of the stockbroker, whom Milgram calls Mr Jacobs. This result was quite mysterious: how was it that so many paths leading to the stockbroker passed through this person?

Frequent air travellers are familiar with a very similar phenomenon. Airports like Heathrow, Frankfurt or JFK in New York are well known to globetrotters: whatever the destination, it is quite likely that flights stop over at those airports. Airline magazines often carry a map of the world, crossed by long lines that show their routes: many of them end up at or pass through places like London, Frankfurt or New York. Airports like these are called *hubs* and carry a large portion of the overall traffic.

It's easy to argue that the role of Mr Jacobs in the social network is the same as the big airports in the air traffic network. Probably, Jacobs was a social relations hub: his many contacts connected him to several people, so it was natural that many letters passed through his hands.

Another striking observation made by Milgram is that a large part of the remaining letters arrived from just two other people: Mr Jones and Mr Brown. Using the air traffic metaphor, these two people were most probably 'average-size airports' (like Madrid or Milan) of the social network. The remaining letters, that didn't came from Jacobs, Jones, or Brown, passed through smaller 'airports' (like Girona or Olbia) of the social network.

The presence of these hubs is not specific to the stockbroker's social network, or to the airport network. Many other systems, when represented as graphs, show similar highly connected vertices, or *superconnectors*. In many networks, one can see a 'winner takes all' tendency: a few nodes attract the majority of connections, and the large remainder of nodes have to share the remaining links. A modern analysis has shown that Mozart's Don Giovanni (who seduced 2,065 women, according to Da Ponte's libretto: '... 640 in Italy, 231 in Germany, 100 in France, 91 in Turkey, but in Spain they are already 1,003...') was not an exaggeration: the most connected individuals in sexual interaction networks can reach thousands of intercourses. In some datasets, some of these are people involved in the sex trade. Naturally, these highly connected individuals are those most needing protection against sexually transmitted diseases. Another example of superconnector was found early after the September 11 terrorist attack on New York: management consultant Valdis Krebs drew a simple map of the social networks of the terrorists and found that Mohammad Atta, one of the leaders of the conspiracy, was the most connected node, that is, a hub. In scientific collaboration networks, one can also find key figures that cooperate with a large number of colleagues: Paul Erdős was one of them.

Superconnectors are present in many kinds of network, not only in social ones. Some routers in the Internet have thousands of connections: that is, thousands more than the average router, which has just a few links. *Meet-me rooms* are large facilities—usually buildings full of cables—where hundreds of Internet Service Providers can link to each other: the failure of one of these facilities can leave entire areas as (as large as a state) without Internet connection. Websites of large newspapers attract an enormous number of links from other websites, blogs, and social networks. In foodwebs, top species predate a large quantity of other species. Finally in words' networks, the hubs are the ambiguous or polysemic words: such as 'arms', which in English refers both to bodily extremities and weapons, and thus connects to a larger semantic or synonym field.

Hubs are also present in the networks inside the cell. In the genetic regulatory network, a single gene can control the expression of a large portion of the rest of the genome: in one bacterium (the *Caulobacter crescentus*), one single regulatory factor (the CtrA) controls 26 per cent of the cell cycle-regulated genes. The p53 molecule is a superconnector of the protein interaction network: the gene associated to this protein is a powerful tumor suppressor, and it is mutated in a large range of tumors. A clear hub of the metabolic network is the Atp molecule (adenosine triphosphate): this plays the role of energy carrier for a large number of biochemical reactions.

Giants, dwarfs, and networks

> And there came out from the camp of the Philistines a champion named Goliath of Gath, whose height was six cubits and a span. [...] He was armed with a coat of mail, and the weight of the coat was five thousand shekels of bronze [...]

> The shaft of his spear was like a weaver's beam, and his spear's head weighed six hundred shekels of iron (1 Samuel 17 4:7)

According to the biblical book of Samuel, Israelites have to wait 40 days before someone dared to face somebody as imposing as Goliath: then enters David, a brave and reckless boy, who ends up defeating the enemy. This was not a common adversary: the 'six cubits and a span' height corresponds to about 3 metres, and the 'five thousand shekels of bronze' weight of his coat of mail would be between 60 and 90 kg, according to historians' estimates.

The conversion of ancient measures into modern ones is not precise; moreover, the biblical narration is most probably symbolic. However, Goliath's size is not completely unlikely. According to the *Guinness Book of Records*, the tallest person ever recorded, an American called Robert Wadlow, was 2.75 metres tall. Unlike Goliath, who has special armour and a spear suited to his size, the exceptionally tall are usually surrounded by objects too small for them: chairs are uncomfortable, ceilings are too low, and they need to wear specially tailored shoes and clothing.

The root of their problems is that body size is a *homogeneous* magnitude. People entering a cinema are of different sizes, but all the seats are identical: some people find them large, others small, but in general they are reasonably comfortable. Body size does not vary a lot from the average. Very tall (or very short) people are exceptionally rare, and the taller (or shorter) they are, the rarer. Almost everybody knows somebody about 1.90 metres tall, but few know a 2-metre tall person and almost nobody a 2.5-metre one. People are also homogeneous with respect to other features. For example, IQ tests yield results close to the average most of the times, and deviations—both upwards and downwards—are rare. Behaviours can be quite homogeneous, too. For example, drivers can be more or less reckless, but most of the time the speed measured on highways conforms quite closely to the average.

However, homogeneity is not always the rule. For example, the number of friends a person can have is extremely variable. Robert Wadlow was 'only' five times taller than the shortest person ever, Chandra Bahadur Dangi, 55 cm tall, according to the *Guinness Book of Records*. By contrast, the friendliest people (the hubs of social networks) can have tens or hundreds of friends more than the extremely shy, who interact with very few individuals. If contacts in virtual social networks are taken as a proxy of the number of friends a person has, then the hubs of these networks have hundreds more friends than the less connected people. While height is a homogeneous magnitude, the number of social connection is a *heterogeneous* one.

If the height of people reflected the number of their social connections, somebody as tall as Wadlow would not enter any record book. There would be people hundreds of times taller than the shorter ones: 'social giants' more than 2 km tall would walk the streets. Even more interestingly, these giants would not be astounding exceptions in a generally short population. All the intermediate heights between dwarf and giant would be represented by some individuals: naturally, the greater the height, the fewer the people of that height; however, the number of tall people in this imaginary world would not diminish as quickly in terms of height as in the real world. In other words, the taller, the rarer: but not as rare as in the real world.

The business of seat manufacturers would be much more difficult in this world, because there would be no way to build a seat that would fit every body size. In the real world, if we want to manufacture seats, or analyse IQ tests, or predict the duration of a road trip, we take into account average height, IQ, or speed. But in order to understand social relations, the very concept of average may be useless. Body size, IQ, road speed, and other magnitudes have a *characteristic scale*: that is, an average value that in the large majority of cases is a rough predictor of the actual value that one will find. In contrast, social relations do not

have this scale. If you knock on the door of an unknown neighbour, you expect to see a person whose height is within a certain reasonable range, and most of the time your guess will be accurate. But it is almost impossible to guess in advance whether that person has many or few friends, and how many. The average number of relationships in a town just gives an idea about whether the social network of that place is more or less dense. But it does not allow us to make any reasonable prediction about each single person.
A system with this feature is said to be *scale-free* or *scale-invariant*, in the sense that it does not have a characteristic scale. This can be rephrased by saying that the individual *fluctuations* with respect to the average are too large for us to make a correct prediction.

Fat tails

In general, a network with heterogeneous connectivity has a set of clear hubs. When a graph is small, it is easy to find whether its connectivity is homogeneous or heterogeneous (Figure 8). In the first case, all the nodes have more or less the same connectivity, while in the latter it is easy to spot a few hubs. But when the network to be studied is very big (like the Internet, the Web, metabolic networks, and many others) things are not so easy.

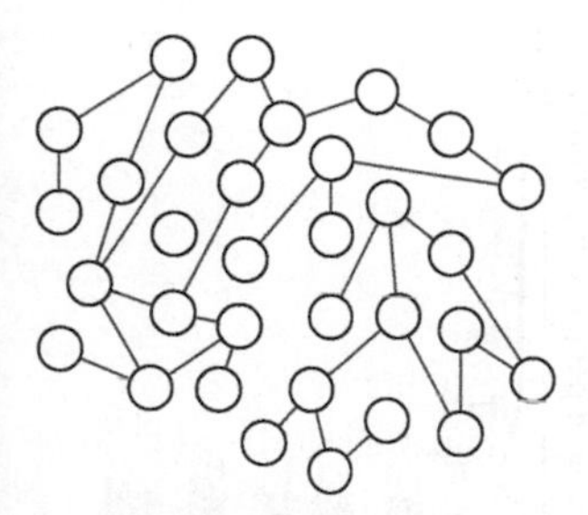

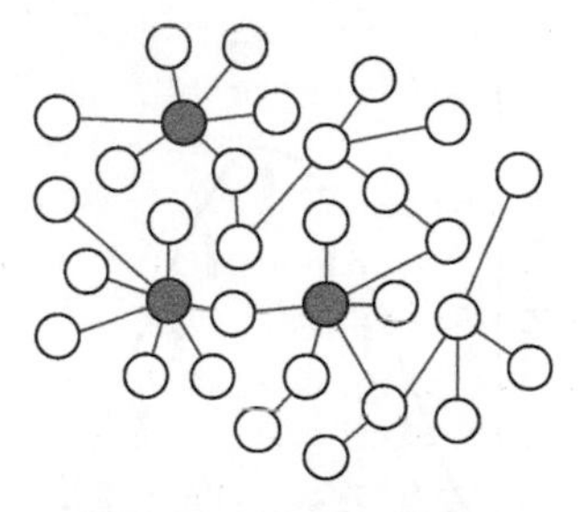

8. A homogeneous network (left), where all nodes have more or less the same degree, compared with a heterogeneous one (right), where highly connected nodes (hubs) are present

Fortunately, mathematics provides a way to find whether a magnitude is heterogeneous or homogeneous.

Let us start with a homogeneous magnitude, such as people's height. In order to study the height of the students of a class, one can do as follows. First, make a row with those 1.50 to 1.55 metres tall: there will be a few of them. Then, make a parallel row with those 1.55 to 1.60: there will be some more and the row will be a little longer. Follow those with 1.60 to 1.65: more people will be in that row. Increasing by 5 centimetres in every row (Figure 9 left). At the end, the profile of the rows will have the shape of a *bell curve*: the number of students increases as height increases, then reaches the top around the average, and then starts to fall. The very tall and the very short are rare and the majority are in the middle. This curve provides the distribution of heights of the students.

Now, let us consider the number of social contacts of those same students. Now the rows correspond to those with 0 to 20 friends, 20 to 40, 40 to 60, and so on. The outcome of this procedure provides the distribution of the connectivity of the nodes of the social network, that is, the *degree distribution* of the graph. The resulting picture is very different from the case of heights (Figure 9 right). First of all, there will be many more rows, since

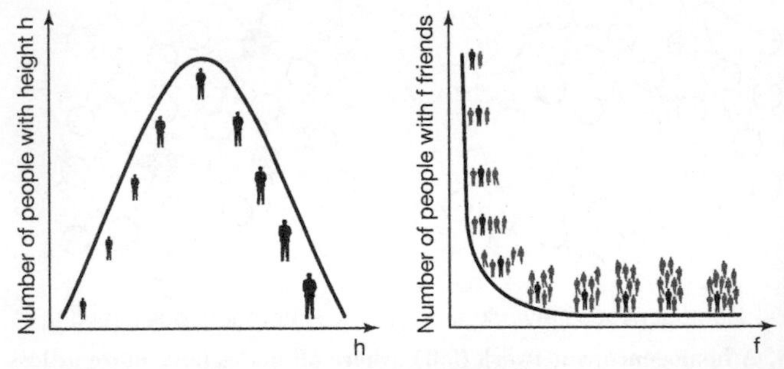

9. Height is a homogeneous magnitude, distributed according to a bell curve (left), while the number of friends is a heterogeneous magnitude, distributed according to a power law (right)

there will be people with hundreds or thousands of friends. The majority of people will have some tens of contacts per person, but the resulting distribution will have a 'fat tail'. In other words, the distribution will be very skewed to the right, with a long or 'heavy tail'. Mathematically speaking, the shape of the degree distribution is well described by a curve called *power law*.

In homogeneous networks, the degree distribution is a bell curve, similar to that of height, while in heterogeneous networks, it is a power law, similar to that of friendships. The power law implies that there are many more hubs (and much more connected) in heterogeneous networks than in homogeneous ones. Moreover, hubs are not isolated exceptions: there is a full hierarchy of nodes, each of them being a hub compared with the less connected ones. Take once again the comparison of height and friendships. Probably there are several million people in the world who are 1.50 metres tall; however, if we double that height (3.00 metres), the number of people this tall is much smaller, most probably zero. On the other hand, tens of millions of people have, say, 20 friends in their social network. If we double that number (40 friends), there will be fewer (say a quarter of those with 20), but still millions. We can double the number many times, and the number will be reduced by about a quarter at every step (the actual rate of reduction depends on the slope of the power law). This explains, for example, the role of Mr Jones and Mr Brown in Milgram's experiment: while Jacobs is the largest hub of the stockbroker's social network, Jones and Brown are smaller hubs, but still well connected.

Looking at the degree distribution is the best way to check if a network is heterogeneous or not: if the distribution is fat tailed, then the network will have hubs and heterogeneity. A mathematically perfect power law is never found, because this would imply the existence of hubs with an infinite number of connections. However, no real network is infinitely big: this is why the fat tail of the degree distribution always has a cut-off at a maximum value for the degree. Indeed, the size of hubs can be limited by various costs of accumulating connections: for example, neurons cannot accumulate an arbitrary amount of connections, because of their physical structure. In professional collaboration networks, time plays a role: connections cannot be accumulated indefinitely, because at a certain moment the career (or life) of an individual comes to an end. All these and other factors are reflected in the shape of the degree distribution. Nonetheless, a strongly skewed, fat-tailed distribution is a clear signal of heterogeneity, even if it is never a perfect power law.

One has to be careful in interpreting what hubs and fat tails mean. For example, some anthropologists believe that a magnitude called *Dunbar number* limits the number of social relations. According to this hypothesis the number of stable social bonds cannot increase much above 150. Anthropologist Robin Dunbar put forward this hypothesis in 1992 after finding evidence that the size of a part of the brain's cortex of primates and humans may be related to that of their social groups. If this is true, what is the explanation of the hubs with thousands of connections found in many social networks? Some scientists think that they are instances of the *pizza delivery guy problem*. A pizza delivery guy receives many phone calls on his mobile, but just a tiny fraction of them come from real friends; the rest are clients. According to this concept, the majority of the links represented in the fat tail of the distribution would be fictitious in social networks. However, this depends on exactly what problem one wants to study. For example, if the pizza delivery guy catches flu, epidemiologists only care about how many people (friends or not) have been in contact with him.

On the other hand, not all networks are necessarily heterogeneous. While the small-world property is something intrinsic to networked structures, hubs are not present in all kind of networks. For example, power grids usually have very few of them. The same holds for some foodwebs, the neuronal network of the *C. elegans*, and the World Trade Web.

Finally, an interesting case is found in some directed networks, as in the case of, most genetic regulatory networks. If gene A regulates gene B, an arrow is drawn from A to B, but not necessarily from B to A. The *out-degree distribution* (that is, the distribution of the number of outcoming arrows) is usually fat tailed: that is, a few genes regulate large portions of the genome. However, the *in-degree distribution* (that of the number of incoming arrows) is much more homogeneous: just a few other genes regulate a single one. Heterogeneity is widespread in many networks, but when we approach an unknown system we must not take it for granted until we have checked it.

The signature of self-organization

Heterogeneity and the lack of characteristic scale may very well be a sign of disorder. One could reason as follows. Many networks (like the Internet or social networks) have grown without any blueprint or supervision. As a consequence, every node in the network follows its own criteria and performs completely different and uncoordinated behaviours. These are so disordered that they can be easily assimilated to an overall random process. As a consequence, random graphs should be good models for these networks. This line of reasoning seems to work, but some problems appear after deeper inspection. The most noticeable is that random networks are not heterogeneous at all. On the contrary, their degree distribution is bell shaped, suggesting that all nodes have more or less the same degree. The process of connecting pairs at random is such that every node ends up with more or less the same degree. More precisely, the degree has a characteristic scale, with small fluctuations around the average. In contrast with many real-world networks, hubs are not present in random networks.

A consequence of this is that, while random networks are small worlds, heterogeneous ones are *ultra-small worlds*. That is, the distance between their vertices is relatively smaller than in their random counterparts. If one takes a random network and adds a certain number of hubs to it (thus making it more heterogeneous), then the distance becomes smaller. Conversely, if one takes a heterogeneous network and randomizes it (that is, builds a network with the same number of nodes and edges, but with edges distributed at random), then hubs disappear, and the average distance becomes larger. This shows that hubs are responsible for the majority of the connectivity of these networks: a large portion of the connections arise precisely from this small number of superconnected nodes.

More importantly, the fact that random networks are homogeneous means that the equivalence between heterogeneity and disorder is flawed. A disordered process such as the one described by random networks does not yield the heterogeneous connectivity found in many real-world networks. On the contrary, heterogeneity may arise from the exact opposite: that is, from some kind of regular, ordered behaviour.

This is rather puzzling, since many networks are not the result of a blueprint, nor do they evolve under tight top-down supervision. Few networks, like the electrical or road ones, are controlled by political and technical authorities, but most of them are unsupervised. The Internet, for example, is controlled by network administrators at the local level, and is also constrained by technical, economic, and geographic features. However, its large-scale structure is largely unplanned: the Internet is very similar to a global-scale experiment, where nobody draws the overall structure, which is the result of the actions of innumerable agents. Biological networks are an even clearer example: there is no designer, only the tinkering effects of evolution. With respect to social networks, politics, money, religion, language, and culture influence the relations between individuals, but when spaces of freedom are available, the shaping of these networks is not strictly planned. In all these cases, the overall organization of the systems emerges from the collective action of its elements, a bottom-up process of *self-organization*. This process may explain why many networks, even without being blueprinted, still display a remarkable signature of order like heterogeneity.

Heterogeneity is not exclusive to networked systems. For example, the intensity of earthquakes has a fat-tailed distribution: if one plots the frequency of earthquakes versus their intensity, a nice power law emerges. The 'average earthquake' does not exist; there is wide variability, from imperceptible vibration to large-scale catastrophes. Another example is the size of cities: this ranges from the largest Chinese megalopolis to small towns in the countryside of Tuscany. Another example is the distribution of income: At the beginning of the 20th century, economist Vilfredo Pareto showed that 80 per cent of Italian land was in the hands of 20 per cent of the population; various levels of skew of this kind are present in all economies.

All these examples share with networks one basic feature: they are the outcome of a complex, largely unsupervised process. Heterogeneity is not equivalent to randomness. On the contrary, it can be the signature of a hidden order, not imposed by a top-down project, but generated by the elements of the system. The presence of this feature in widely different networks suggests that some common underlying mechanism may be at work in many of them. Understanding the origin of this self-organized order is one of the central challenges of the science of networks.

Chapter 6

Emergence of networks

Everlasting change

By the nineties, the Internet was mostly an unknown land. Although it was already a critical infrastructure for communication, trading, and transportation, nobody had a clear idea about its overall architecture. Administrators controlled local networks, but had no clue about the large-scale structure of the net. Moreover, the increase had been explosive: from some tens of machines at the beginning of the seventies, to tens of millions, with even larger prospects of growth. By the end of the decade, organizations like the computer company Compaq and the Cooperative Association for Internet Data Analysis (CAIDA) launched a series of *mapping projects*, aimed at exploring the Internet and drawing its global layout. Compaq's project was called *Mercator*, in honour of the geographer that drew in the 16th century one of the the most important maps of the world, including the recently discovered America: the Internet was acknowledged to be a 'new world' to be explored. Thanks to these and other projects, maps of the Internet are available, and its growth is now monitored.

The dynamics of the Internet have not stopped: at this very moment, routers, computers, cables, and satellite connections are constantly being added or removed, with the net effect of continuous growth. Despite the absence of a plan, this is not a completely random process. On the contrary, the Internet is a highly ordered and efficient structure. This emerging order must be the result of some regularity in the behaviour of individual agents that build the Net. There must be some small-scale mechanism that, iterated through a great number of interactions, ends up generating a structure that is organized at the large-scale level. Disentangling the general principles underlying those processes that shape network structures is essential to understanding their self-organization.

Even networks seeming to be completely static are subject in fact to some kind of dynamic process. The networks of genes, proteins, and metabolites in the cell, those of neurons in the brain, and those of species in the ecosystem, seem to be fixed in time: genetic regulation, metabolic pathways, neuron connections, or prey–predator relations are relatively stable. However, the networks within the cell live an explosive growth during the development of each individual and are constantly changing as the organism ages, and in reaction to the environment. The plasticity of the brain may decrease throughout life, but it is never completely lost. Extinctions or invasions of new species radically reshape foodwebs. Moreover, all biological nets change in the long run, for the action of natural evolution. Something similar happens in other apparently static networks, such as power grids or language networks. After power grids are set up, they slowly modify their shape due to accidents and technological evolution. The network of words changes as speakers change, and as the overall language evolves, introducing neologisms and new semantic relations.

In other networks, the change is mostly concentrated in the connections, while the set of vertices is almost immutable. For example, every day the banks of a given country set up a different pattern of money lending, in the interbank network. Occasionally, the set of vertices of this network can change, for example if a bank goes bankrupt or if a new one appears on the market. However, this change happens on a much longer timescale than the transactions (i.e. the rearrangement of edges): on a given day, the changes in the networks are mainly associated to the edges. Something similar happens in the networks of price correlations between stocks (correlation varies much more frequently than the actual set of stocks), in the World Trade Web (changes in economic relations between countries of the world are more common than the creation of new countries, through separation or federation), and in the airport network (in any one year, only a few new airports open, while a great number of flight connections change).

Exactly the opposite happens in another group of networks: in these graphs, new nodes are constantly added, and this process is much more relevant than the rearrangement of links. The most exact realization of this case is the network of citations within scientific papers. New papers appear every day, with citations to older ones, and once they are published, the respective citations cannot be changed any more.

Finally, there are networks where the dynamics of adding (or eliminating) nodes and connections happen at the same rate, giving place to a quite complicated process. The Web is constantly updated with creation and deletion of both new web pages and new hyperlinks. On some specific websites, such as Wikipedia, new articles and new connections between articles are created every day.

The range of possible network dynamics is extremely wide: network scientists have made measures, mathematical models, and computer simulations in order to grasp the basic mechanisms underlying this process, in the hope of understanding the principles of self-organization of these networks.

The rich get richer

At the beginning of the sixties, sociologist Harriet Zuckerman interviewed a number of scientists who had received the Nobel Prize. She aimed at finding what was so special in their way of working, and what the secret of their success in research was. She found a recurring theme in the answers of the Nobelists. A laureate in physics said: 'The world is peculiar in this matter of how it gives credit. It tends to give the credit to [already] famous people.' A laureate in chemistry added: 'When people see my name on a paper, they are apt to remember *it*, and not to remember the other names.' And a laureate in physiology and medicine specified: '[When you read a scientific paper] you usually notice the name that you're familiar with. Even if it's last it will be the one that sticks [. . .] You remember that, rather than the long list of authors.' 'The man who's best known gets more credit, an inordinate amount of credit,' concluded a laureate in physics.

Those observations led another sociologist (and later husband of Zuckerman), Robert Merton, to formulate in 1968 a brilliant law. Merton put forward that science is under the effect of a social mechanism dubbed the *Matthew effect*. The name comes from the Gospel of Matthew:

> For to all those who have, more will be given, and they will have in abundance; but from those who have nothing, even what they have will be taken away (Mt, 25:29).

According to Merton, this mechanism plays a role in the distribution of prizes, financing, visibility, prestige . . . Scientists with a good number of these assets acquire even more of them with ease. On the other hand, those without them have a hard time acquiring and retaining them.

In 1976, physicist and science historian Derek de Solla Price found quantitative evidence that supported this view. Price analysed a large set of scientific papers, linked by mutual citations. He found that papers with a good number of citations at a particular time tended to acquire even more citations later, while those that had only a few at the beginning did not increase their citations that much. Price showed mathematically that this simple rule could explain the appearance of highly cited papers (hubs of the citation network). More precisely, he showed that this mechanism could explain why the distribution of the number of citations per paper displayed the characteristic fat tail of a power law.

Price's model was a variant of mathematical models developed earlier by statistician George Udny Yule and social scientist Herbert A. Simon, but it was not until 1999 that it became clear that this mechanism could explain the appearance of hubs, heterogeneity, scale-invariance, in synthesis of fat-tailed distributions, in a set of different networks. This realization is due to physicists Albert-László Barabási and Réka Albert. Barabási and Albert put forward a mathematical model of the growth of a network. They imagined a graph that starts with a small set of vertices (even just two or three), connected at random. New nodes are added at a steady rate to this initial nucleus, each of them carrying a given number of links. A simple rule establishes how new nodes are linked: incoming vertices prefer old ones that already have many links. This mechanism is called *preferential attachment* (Figure 10). In principle, new vertices can attach to any of the old ones, but the higher the degree of an old node, the higher the probability of attracting a new one. Occasionally, less connected nodes will receive new links, but most of the time hubs will be much more attractive.

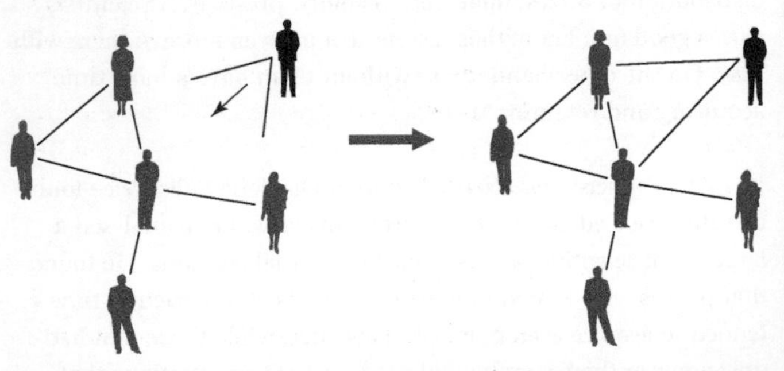

10. In the preferential attachment mechanism for network growth, new nodes connect preferentially with old nodes that have a high degree

This process can be studied mathematically and simulated on computer, and one can take measurements in real networks to see whether it is at work or not. At the beginning, all the nodes have more or less the same degree. However, during growth, some of them start to accumulate more links than others. The more connections a node has at a certain moment, the more it is capable of attracting new connections. This is why the preferential attachment is also called the *rich-get-richer* mechanism. As a consequence, the initially small differences in connectivity are progressively amplified. Thus, a hierarchy of different nodes emerges, with a large variety in their degrees, ranging from the least connected ones to those that have accumulated many links, i.e. the hubs. The resulting network is heterogeneous, with a power-law degree distribution.

The Barabási–Albert model shows that a bottom-up mechanism of growth can generate heterogeneity, without imposing any top-down blueprint. The global scale-invariance of the network is the outcome of the iteration of an individual, local choice: preferring more connected nodes to less connected ones. The model uses probability to allow for individual deviations from this behaviour: some nodes can decide to connect to low-degree nodes. However, the general tendency sets the outcome. As a further confirmation, one can check that the heterogeneity of the network disappears if it is grown without the preferential attachment rule. Indeed, new nodes connect to old ones at random, in such a way that the degree of old nodes does not influence their capacity to attract new links; the outcome is a homogeneous network in which every node ends up having more or less the same degree.

A widespread mechanism

Preferential attachment is the network version of a mechanism at work in many natural and social phenomena whose future evolution depends on their history. For example, the size of cities changes in time according to their present size: large cities experience large increases, small cities have small changes. Tomorrow's stock prices are also proportional, on average, to today's prices. This mechanism is also called *multiplicative noise*. There are various reasons why such a process may be at work in many networks. In some situations, having many links is the main way to be discovered by new nodes. A website linked to by many others is more easily found by people browsing the Web than is a less linked one. The same happens to highly cited papers. This increased visibility makes it easier to receive even more links or citations.

In other cases, links are attractive for their own sake. Sociologists have found evidence of *indirect mate choice* in which people choose their partner not only on the basis of that individual's personal features, but taking into account as well other people's opinion. For example, a study found that college-age women tend to rate higher a man in a picture where he is surrounded by many other women than the same man photographed alone. A long list of previous partnerships may put a person in a good position to attract even more partners. This is why preferential attachment is also called the *popularity is attractive* principle.

A more conscious reason for preferential attachment is that linking to hubs grants easier access to many other nodes. Since deregulation, in 1978, many airlines have adopted a *hub policy*, consisting in choosing well connected airports as their favourite destinations. The incentive is clear: since these airports provide access to a large range of destinations, connecting to them attracts more potential clients. Something similar happens in the web of interlocking directorates: a director that sits on many boards has access to a lot of information and a broad vision, something that makes it very attractive to hire him or her onto even more boards. Access is essential in the Internet, too. A large part of the Internet is set up and maintained by private companies, called *Internet Service Providers* (ISPs). When one of them sets up a new infrastructure, its priority is granting users rapid access to the information stored in the network. With this idea in mind, ISPs do not choose at random the routers to which they want to connect, nor do they just choose the closer ones. On the contrary, they select nodes that grant access to the greatest possible number of servers, in the smallest possible number of steps. And what is better than a hub to achieve this? In the case of the Internet, the data provided by the mapping projects seem to agree with the preferential attachment hypothesis. Figures show that vertices with many connections in a map issued at a certain time tend to have even more connections in the next map.

In some cases, a preferential attachment process comes in the guise of other mechanisms. Imagine someone wants to create a personal web page. A common strategy is to look at friends' pages and pick a nice one to use as a template. Since a person usually shares common interests with his or her friends, the new web page will likely keep most of the links of the template, maybe changing a few of them. So in the end, the new page will be a copy of the template, with a few changes. Now, this mechanism disguises a form of preferential attachment. In fact, to what pages does the template point to? Most likely to hubs. Only because hubs exist, and capture a large fraction of the connectivity of the WWW, any given page will point more likely to hubs than to weakly connected pages. So, if a page is duplicated, each copy brings even more links to hubs, resulting in an effective rich-get-richer process, so that the preferential attachment rule is recovered. The copying mechanism seems weird, but in fact it is a leading factor in several situations. For example, scientific citations included in an article are often drawn from other articles in the same field, and tend to consolidate the authorities of the sector.

But one of the most interesting applications is in genetic networks. Genomes often evolve through the process of *duplication and diversification*. During cell replications, all the DNA is copied to the new cells, but sometimes a mistake happens: a full gene of the original DNA chain is duplicated, and appears twice in the genome of the daughter cell (*duplication*). Most of the time, this new gene just produces redundant proteins that do the same as its twin. However, in further replications, one of the two can suffer a mutation that may allow its protein to perform new biological functions, for example interacting with different proteins than its former twin (*diversification*). This evolutionary mechanism has been observed in many cases. Now, its translation in terms of the protein interaction network is totally akin to a copying mechanism: a new node enters the network (the copied and mutated protein), with some of the links of its ancestor (the proteins with which it originally interacted), and some new ones, due to the mutation. Highly connected proteins have a natural advantage in this mechanism: it is not that they are more likely to be duplicated, but they are more likely to have a link to a duplicated protein than the weakly connected ones, and therefore they are more likely to gain new links.

Although the role of gene duplication has been shown only for protein interaction networks, there is evidence in favour of preferential attachment (either direct or in disguise) as well in metabolic networks. An obvious consequence of this linking procedure is that usually hubs are between the older nodes in the network, because they have had the opportunity to profit from 'first mover advantage'. Now, metabolic hubs are precisely primitive molecules, possibly incorporated in the genome during the evolution of the first forms of life, remnants of the RNA world such as coenzyme A, NAD, and GTP, or elements of the most ancient metabolic pathways, such as glycolysis and the tricarboxylic acid cycle. In the context of the protein interaction networks, cross-genome comparisons have found that, on average, the evolutionarily older proteins have more links to other proteins than their younger counterparts.

Preferential attachment is not the only mechanism at work in networks, and not all heterogeneous networks come from this mechanism. However, the Barabási–Albert model gives an important take-home message. A simple, local behaviour, iterated through many interactions, can give rise to complex structures. This arises without any overall blueprint, and it appears even if we allow for a certain level of randomness in the behaviour, accounting for individual deviation with respect to general trends.

When fitness matters

When an occasional sexual relation is at stake, people tend to be very tolerant with respect to certain features of the potential partner, such as political ideas, social class, or whether he or she smokes. However, these elements become very important when considering engagement or marriage. This is the message from an analysis of mid-nineties data by sociologist Edward O. Lauman. These data show that about three-quarters of the married couples in the US share a wide range of similar traits. These include belonging to the same social class, ethnic group, and education level, and even sharing similar levels of attractiveness, political ideas, and health behaviours, such as eating and smoking habits. On the other hand, when other types of sexual relationships are at stake, the proportion is much lower, although still high (more than half).

Homogamy, the tendency of like to marry like, is very strong even in societies where marriages are not combined and, theoretically, everybody could match with everybody else. In the turbulent college years, the prestige of an individual—as measured, for example, by the number of his or her previous sexual partnerships—can be a relevant driving force in shaping the sexual relations networks. But when people settle down, much stricter criteria come into force. While in the first case the rich-get-richer process can be at work, it can hardly explain the second case. Homogamy is a specific instance of *homophily*: this consists of a general trend of like to link to like, and is a powerful force in shaping social networks, according to much sociological evidence. In the Barabási–Albert model, the main criterion for linking to a node is the number of its links, but in many situations, other features, independent of the actual number of links, are much more important for attracting new links.

A consequence of the Barabási–Albert dynamics is that old nodes have a cumulative advantage over new ones. However, this is not always the case in practice. For example, old glories of the Web, such as Magellan or Excite search engines, are now mostly forgotten. Newer ones, such as Google or Yahoo!, have taken their place. Cumulative advantage can be completely overthrown when a new actor enters the game (think Facebook, for example). Newcomers often have some intrinsic feature that makes them much more attractive than older players. In this case, the connectivity of a network is not exclusively driven by the degree of the nodes, as in the Barabási–Albert model. On the contrary, a particular character of each node can play a very important role in its ability to gain links. This character is referred to as the *fitness* of the node, or as its *hidden variable*, a feature that shapes the structure of the network without being as evident as the number of links.

In 2002, physicists Guido Caldarelli, Andrea Capocci, Paolo De Los Rios, and Miguel Ángel Muñoz devised a model to generate a network only on the basis of its nodes' fitness. The basic recipe is identical to the random graph: within a set of nodes, all the possible pairings are considered, and a link is drawn or not between each pairing, according to a given probability. However, in this case, the probability is not fixed, but changes depending on the fitness values of the nodes in the pair. The first step in the model is distributing fitness to the nodes. This hidden variable may represent, for example, the income of an individual. If this is the case, it would be distributed between the nodes in such a way as to mimic the wealth distribution of a country: a few very rich nodes, then a certain number of upper-middle-class nodes, then lower-middle-class, and so on. The second step is defining the probability of linking. In order to simulate a highly segregated society, one could establish the following rule: the probability that two individuals form a social relation depends on the difference between their incomes; specifically, it is inversely proportional to it. That is, the higher the difference of income, the lower the probability they get connected. There is always some chance that two individuals with very different incomes could get connected, but in any case this will not be the leading trend: in general, homophily will triumph.

The fitness model may seem an overly simplified mechanism, but in some cases it works perfectly: for example, in the world trade web. In this case, the fitness is the GDP of the countries of the world. The linking rule is the following: the higher the fitness of two countries, the higher the probability that they get connected: this would be a kind of *fit-get-richer* mechanism, in which the countries with highest GDPs tend to build more commercial links with each other. This is a different mechanism from homophily, because while high-GDP countries establish many links with similar ones, low-GDP countries do not link with other low-GDP countries; they just stay poorly linked. Physicists Diego Garlaschelli and Mariella Loffredo showed in 2004 that these ingredients are enough to predict with great precision the features of the world trade web, provided that one feeds into the model the overall number of existing commercial links at any given time. For example, the model predicts exactly the shape of the degree distribution of the network. This suggests that the basic mechanisms underlying the self-organization of the world trade web are captured by the simple dynamics of the fitness model.

Like preferential attachment, it is unlikely that the fitness mechanism is at work in all real-world networks. While the Barabási–Albert model is plausible when applied to growing networks, the fitness model works also for static networks, where the number of nodes is mainly fixed. Nevertheless, they can be at work at the same time: in 2001, physicists Ginestra Bianconi and Albert-Lászlό Barabási introduced the idea of fitness in a preferential attachment model, showing that a mixture of the two effects gives a good prediction of the Internet's topological properties. It has to be noticed that the fitness model does not always yield a power-law degree distribution. A broad range of distributions of fitness and linking rules generates it, but many others do not. However, this is not a limitation, but rather a positive aspect of the model, that allows it to be applied also to non-heterogeneous networks such as the world trade web.

A variety of strategies

The myth of the girl (or boy) next door has passed into history. In the mid eighties, the number of people that reported meeting their spouses in their own neighbourhood was already negligible, about 3 per cent in France, according to a 1989 study by sociologists Michel Bozon and François Heran. However, just three decades earlier, this case was quite common: Bozon and Heran found proportions of 15 to 20 per cent between 1914 and 1960. In a number of countries in the world, this is still the situation. In some settings, marriages (and social relations in general) are driven neither by some kind of popularity nor by similarity criteria: when geographical constraints matter (for example, if regular access to long-range transport is not available), one is forced to make friends with neighbours and fellow villagers.

In these cases, the vertices of a network are embedded in physical space and this has many important consequences. In some cases, one can connect at almost no cost with anybody else (making friends on a virtual social network). But in other situations, a long-distance connection is quite costly. The properties of the networks in the two cases are then quite different. Many infrastructure networks (trains, gas tubes, highways, etc.) display this bias, since they are embedded in the physical space. Other networks are embedded in time: for example, scientific papers are published on a certain day, and this creates a bias in linking, namely that new papers can cite only older ones, while old papers cannot cite newer ones.

Other biases and strategies can influence network formation. Sociologists have identified two basic incentives for linking in social networks: *opportunity-based antecedents*, that is the likelihood that two people will come into contact, and *benefit-based antecedents*, that is some kind of utility maximization, or discomfort minimization, that leads to tie formation. The global *optimization* of some quantity can play an important role in shaping technological networks: for example, the pressure to minimize search costs on the WWW can lead to a tendency to optimize the shortest path lengths and the link density.

Finally, it may even be that apparent self-organization comes from complete randomness. Imagine that a company releases a new social network and provides nicknames to 100,000 people. Then the company gives permission to one individual to connect with 1,000 others, permission to two individuals to connect with 500, permission to three to connect with 333, permission to four to connect with 250, and so on. People do not know who is behind the nicknames, so they choose their partners at random. Obviously, there is no self-organization in this process: the rules established by the company determine the structure. Still, the final network has *by construction* a power-law degree distribution. This example shows that, in some cases, power laws do not necessarily mean self-organization.

All this large range of strategies, biases, processes, and motivations must be taken into account when trying to understand the features of a network by modelling its behaviour. It may even be that each individual network needs a model of its own. However, it is likely that some very general mechanisms, such as preferential attachment or fitness-related dynamics, play a role in the formation of large classes of apparently unrelated networks. The models described in this chapter are simple explanations of the ways in which local mechanisms, without global planning, can indeed generate large-scale, complex, ordered, and efficient structures.

Chapter 7
Digging deeper into networks

Who are your friends?

For every white American with a sexually transmitted disease, there are up to twenty African-Americans with the same condition in certain areas of the US, according to several studies carried out in the nineties. This figure is the outcome of persisting racial inequalities. However, the actual mechanisms of contagion that generate such a big difference are still partially obscure. In 1999, sociologists Edward O. Laumann and Yoosik Youm found an interesting piece of evidence: less sexually active African-Americans (those who had only one partner in the previous year) were five times more likely to have intercourse with more sexually active African-Americans (those who had four or more partners in the previous year) than whites in the same situation. In other words, in the sexual interaction network of the whites, the *periphery* of less active people was partially separated from the *core* of active individuals. On the contrary, these two groups were more connected in the African-American network. The reason for this difference is unclear, but its consequence is straightforward: in the first network, sexually transmitted diseases thrive mainly within the core, while in the African-American one, they also spill over to the periphery.

This is a case in which the degree of the nodes in the networks is not the most relevant quantity in understanding the situation. Individuals with exactly the same number of sexual connections have a different exposure to infection depending on whether they are African-Americans or whites. In situations like this, it is not enough to know how many 'friends' you have (your degree): it is necessary also to know how many friends your friends have. The degree distribution provides a great deal of information about the overall structure of a graph, for example whether it has hubs or not. However, it does not tell everything about that graph. For example, consider two networks with the same number of nodes and edges: the nodes may have exactly the same degrees, but the edges can be arranged in such a way that the outcomes are very different graphs. The degree is a local feature of a vertex. In order to capture the more subtle structure of networks, one has to dig deeper, and find measures that describe the surroundings of a node: its nearest neighbours, the neighbours of its neighbours, etc.

In the whites' sexual network, low-degree nodes tend to connect with low-degree nodes, and high-degree nodes with high-degree nodes. This phenomenon is called *assortative mixing*: it is a special form of homophily, in which nodes tend to connect with others that are similar to them in the number of connections. By contrast, in the African-Americans' sexual network, high- and low-degree nodes are more connected to each other. This is called *disassortative mixing*. Both cases display a form of correlation in the degrees of neighbouring nodes. When the degrees of neighbours are positively correlated, then the mixing is assortative; when negatively, it is disassortative.

Usually, the presence of these patterns of mixing is the outcome of some non-trivial mechanism acting in the network, possibly a form of self-organization. In random graphs, the neighbours of a given node are chosen completely at random: as a result, there is no clear correlation between the degrees of neighbouring nodes (although the finite size of a graph can disguise this to some extent). On the contrary, correlations are present in most real-world networks. Although there is no general rule, most natural and technological networks tend to be disassortative, while social networks tend to be assortative. For example, highly connected web pages, autonomous systems, species, or metabolites tend to be linked with less connected nodes of their networks. On the other hand, company directors, movie actors, and authors of scientific papers tend to link with those similar to them in connectivity: the higher the degree of an individual, the higher that of his or her neighbours in the network.

Degree assortativity and disassortativity are just an example of the broad range of possible correlations that bias how nodes tie to each other. For example, Laumann and Youm also showed that African-Americans, much more than other groups, tend to have partners from their own community. As a consequence, when an infection enters the community, it gets 'trapped' in it. This simple effect alone makes the likelihood of African-Americans having a sexually transmitted infection 1.3 times greater than the figures for white Americans. In this case, the correlation does not arise from the degree, but is a form of homophily with respect to an intrinsic character of each node, namely its ethnic identification. Another example is correlation with respect to body mass: it has been shown that people with similar body mass index tend to establish social bonds with each other more frequently than with other people. One must note that correlations do not always need to be positive, favouring homophily: for example, in foodwebs, edges connect plants to herbivores, and herbivores to carnivores, but very few connect herbivores with herbivores, or plants with plants.

Who are your friends' friends?

Cosimo de' Medici, the man who led his family to take over Florence in 15th century, was described as an 'indecipherable sphinx'. Although he rarely spoke publicly, and never committed openly to almost any form of action, he was able to build around him a strong party that made him the *pater patriae* (father of the nation) of the most important city of the Renaissance. In 1993, social scientists John F. Padgett and Christopher K. Ansell analysed the information about marriages, economic relations, and patronage links that connected the Medici to the other powerful families of the city. They found that Cosimo's family was at the centre of a network of ties with many of the leading lineages. More importantly, without the Medici connection, most of the time those families were weakly related, or even opposed to each other. Cosimo's reserved attitude helped him to establish relations of alliance and control with everyone.

The network with the Medici at the centre is an example of an *ego network*, a graph composed of a set of nodes with direct ties to a central one (the *ego*), as well as ties linking them to each other. Whenever one of the latter ties is missing (i.e. two neighbours of the ego are not neighbours to each other) the network has a *structural hole*. Cosimo's network was full of structural holes, and his family was able to use them to implement a *divide et impera* (divide and reign) strategy: the Medici were seen as a third party in many conflicts, and families had to ask their mediation in their relations with other families.

However, being surrounded by many structural holes is not always beneficial for an individual. Adolescent girls whose friends are not friends with each oth latter are twice as likely to commit suicide, according to a 2004 study. A possible explanation of this finding is the exposure to conflicting inputs from unrelated friends. Another example comes from workers' unions: when workers' networks lack structural holes (that is, *egos* are surrounded by nodes with abundant mutual ties), then a powerful, well coordinated, and communicative organization arises. In general, different patterns of structural holes point to different situations. For example, a scientist working in a specialized field is usually connected to other scientists in the field, which are likely connected to each other. On the other hand, a scientist working in a highly interdisciplinary field is probably connected to experts of various areas not necessarily in contact with each other.

In all these cases, it doesn't matter how many friends you have (your degree), or who they are (for example, whether their degree is similar to or different from yours). What really matters is who your friends' friends are; in particular whether or not your friends are also friends to each other. This concept is often referred to as *transitivity*, or *clustering*. Let us consider an individual with two friends: they form a *connected triple*. If the two friends are friends with each other, then they also form a *transitive triple*, or *triangle*. The quantity of triangles in a network compared with the overall number of connected triples is the basic ingredient of the *clustering coefficient* of that network: this is a measure of the density of triangles in that graph, its overall transitivity. In random networks, the connections between the nearest neighbours of a node are as random as those between any two nodes. As a consequence, these graphs have just the quantity of triangles that emerge from a purely random disposal of edges. On the other hand, the clustering coefficient of almost all real-world networks is higher than their random counterparts. This suggests that some non-trivial process, possibly a form of self-organization, is at work in generating this extra transitivity.

The high clustering of many networks suggests the presence of groups where 'everybody is friends with everybody else'. At first sight, this picture seems to contradict the small-world property of networks: are networks 'open' worlds in which everybody is a few steps from everybody else, or are they the sum of tightly knit, segregated groups? In reality, there is no real contradiction between these two features: one can see this by having a closer look at the Watts–Strogatz model. The model starts by considering a circle of nodes, each of them connected with its first- and second-nearest neighbours, such as remote villages interchanging goods with their neighbours. This is a fully clustered structure, in which all the commercial partners of one village are commercial partners with each other. The model then allows for rewiring a few links to randomly chosen nodes: a few villages open paths to other faraway villages and bring goods there, declining to do business with one of their neighbours. A few paths are enough to reduce abruptly the distance between any two villages, but on the other hand we can think that the local tight commercial structure is disrupted, that is, its clustering goes down. However, Watts and Strogatz found that the decrease in the clustering is less pronounced than the decrease in the average distance. In practice, in order to make a noticeable drop in the transitivity, one has to rewire almost all the nodes. At this point, only random connections are present in the network. Since it is a random graph, we do not expect a large clustering. The take-home message is that networks (neither ordered lattices nor random graphs), can have both large clustering and small average distance at the same time.

Another interesting point about clustering is that in almost all networks, the clustering of a node depends on the degree of that node. Often, the larger the degree, the smaller the clustering coefficient. Small-degree nodes tend to belong to well-interconnected local communities. Similarly, hubs connect with many nodes that are not directly interconnected. In the Internet, for example, low-degree autonomous systems usually belong to highly clustered regional networks, interconnected by national backbones. A similar structure is likely to be present in many networks, where clustering decreases with increasing degree.

Who are your friends' friends' friends'...?

Money does bring some happiness, but being surrounded by happy people gives much more of it. Earning US$5,000 more per year increases the chance of being happy by just 2 per cent, according to a 1984 estimate, while having a happy friend increases it by 15 per cent, according to a 2008 study by sociologists Nicholas Christakis and James Fowler. The two scientists asked more than 12,000 people from Framingham, Massachusetts about their subjective feeling of happiness. Moreover, they mapped who was a friend, spouse or sibling of whom. By drawing this network, the two found that connected people tend to have similar feelings: happy people tend to group together, as, on the other hand, unhappy people do. Christakis and Fowler found even more interesting evidence. The happiness of an individual is influenced by the happiness of people that are not their immediate neighbours. The 'happiness effect' two steps away (friends of friends) is about 10 per cent; three steps away (friends of friends of friends), it is about 6 per cent. The effect fades only at the fourth step. The two sociologists and other scientists found similar results for obesity, smoking habits, and word-of-mouth advice (such as finding a good piano teacher or finding a home for a pet): in all these cases, influence and information arrived to an individual from three degrees away. This *three degrees rule*, found in several social processes, is an example of *hyperdiadic spread*: that is, the diffusion of a phenomenon beyond *dyadic* relations, those that connect nearest neighbours. In this case, it is neither the degree of a node, nor the degree of its neighbours, nor the connections between its neighbours that matter. The influence goes beyond the immediate circle of each node. Actually in many phenomena, it goes even beyond the third degree. For example, a highly infectious disease can form longer chains of contagion; similarly, the spread of nutrients in a foodweb can span all the network.

In this kind of dynamics, a node can be more or less important depending on the number of chains passing through it. In order to capture this idea, sociologist Linton C. Freeman introduced the concept of *betweenness centrality* of a node. One takes all the pairs of nodes in a network and counts the shortest paths connecting them. The betweenness centrality of a node is basically the proportion of shortest paths that cross that node. The higher this proportion, the more central is the node. According to this measure, the Medici are the most central family in the network of lineages in 15th-century Florence. In this case, the betweenness centrality is a measure of the potential to slow down the flow or to distort what is passed along, in such a way as to serve the node's interests. Several studies show that the centrality of a firm in economic networks predicts well its ability to innovate (as measured by number of patents secured) as well as its financial performance. Interestingly, between 1980 and 2005, East Asian countries experienced huge increases in their centrality in the world trade web, while the centrality of most Latin American countries went down. However, the trade statistics of these two regions displayed similar patterns: the big difference in their development was not tracked well by macroeconomics statistics, while the network-based approach captured it. Moscow became in the Middle Ages the most central node of the river transportation network of central Russia, according to a 1965 study. Very probably this set the scene for the future importance of the city.

Central nodes usually act as bridges or bottlenecks: they are almost compulsory stops in the traffic on a network. For this reason, centrality is an estimate of the load handled by a node of a network, assuming that most of the traffic passes through the shortest paths (this is not always the case, but it is a good approximation). For the same reason, damaging central nodes (for example, extinguishing a central species or destroying a central router) can impair radically the flow of a network. Depending on the process one wants to study, other definitions of centrality can be introduced. For example, *closeness centrality* computes the distance of a node to all others, and *reach centrality* factors in the portion of all nodes that can be reached in one step, two steps, three steps, and so on. More complicated definitions are also available.

The behaviour of centrality in many real-world networks is a further signature of their heterogeneity. In many real-world networks it exhibits the characteristic long tail of a heterogeneous distribution. The average centrality is not a valid estimate for the centrality of any node, because this magnitude varies a lot around the average: a few nodes are the main bottlenecks of almost all the shortest paths in the network, and a full hierarchy of less central nodes goes down from them. Given the importance of the more central nodes, it is natural to ask whether they are the same as the hubs of the networks. In many situations, this is in fact the case. For example, highly connected autonomous systems also act as bridges between regional networks; or polysemic words, with their many connections to other words, bring together separated areas of language. But this is not a general rule. A notable exception is the airport network: in this case, certain low-degree airports have exceptionally large betweenness. In 2000, the airport with highest centrality was Paris, a hub connected to more than 250 other cities. But the next highest was remote Anchorage, in Alaska, a medium-sized airport with just 40 connections. Other airports similar to it appear in the list of the most central ones. How is this anomaly explained? Alaska has many airports for internal flights, but Anchorage is its only bridge to the rest of the US, so many paths cross this airport. The anomaly is the result of the existence of regions with a high density of airports but few connections to the outside world.

What group(s) do you belong to?

In 1972, two karate instructors in a university club in the US were so much in conflict that they decided to split their club into two different ones. This event, pretty unexciting for most of the world, was a goldmine in the eyes of social scientist Wayne W. Zachary. In 1977, he published a pioneering study in which he gave an unconventional view of this event.

In 1970, the karate instructor, Mr Hi, asked the president of the club, John A., to raise the prices of lessons in order to provide a better salary. All he received was a denial. As time passed, the entire club became divided over this issue, and after two years the supporters of Mr Hi formed a new organization under his leadership. During all that time, Zachary collected information about the karate lessons, the meetings, the parties, and the banquets of club members, and identified as good friends those that also met outside the club. At this point, he was able to draw a precise network of the friendships within the club: the resulting structure was clearly divided into two groups built around the two instructors, each of them composed of people who were friends with each other and with one of the instructors, with few connections with people in the other group (Figure 11). When the club split in two, the people divided almost exactly along the lines that separated the two groups.

Zachary's method was able to predict the club division almost perfectly, on the basis of the structure of the network alone. Since then, researchers have been striving to find a general method to identify *communities* or *modules* in networks. In Zachary's case, they could be seen just by examining the map, but other instances are much more complicated, and a general solution is still to be

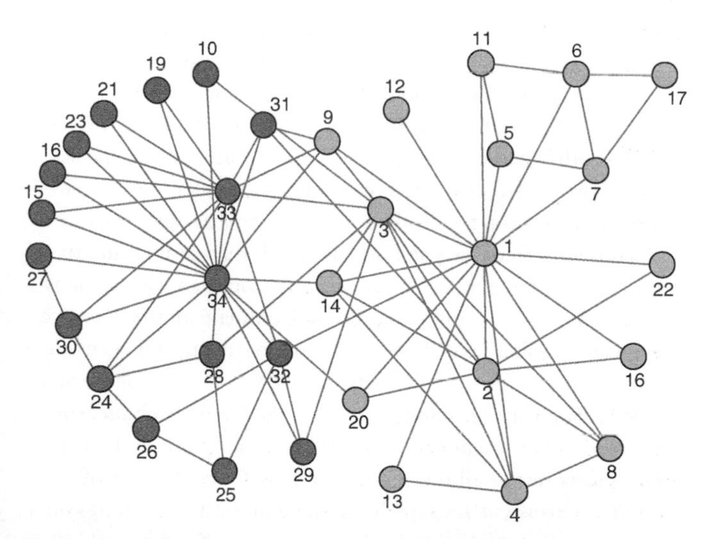

11. The structure of friendships of a karate club studied by anthropologist Wayne Zachary allows us to predict the separation of the group into two communities

found. All real-world networks display some level of modularity. Alaska is obviously a specific module within the airport network structure, such as other regions well connected to the inside but not to the outside. Foodwebs are divided into *compartments*: groups of species interacting more strongly among themselves than with others. Social networks are divided into *cliques*: for example, studies with adolescents have showed that their behaviours are strongly influenced by the modules to which they belong. The neuronal network is divided into big areas, often corresponding to specific functions. The genetic regulatory network is divided into subnetworks, associated to specific functions or diseases. Degree, correlations, clustering, and centrality provide information on single nodes, their immediate surrounding, and their position with respect to the overall network, but they do not capture the discrete structures into which the overall graph is divided.

The simpler form of module is the *motif*, a pattern of connections within a few nodes repeated throughout the network. In foodwebs we frequently find a diamond-like structure: for example, a carnivore eats two different herbivores, and they both eat the same plant. Another common motif is a simple chain of three species: a big fish eats a small fish that eats an even smaller one. These patterns are not the result of pure chance: motifs appear with a much higher frequency in a real foodweb than in its random counterpart. In general, in large networks, one can isolate many subsets of nodes and edges that may be candidate motifs. However, a given subgraph is considered a relevant motif only if it occurs in a network with a higher probability than in its random counterpart. In the Web, a very common example is the *bipartite clique*: this is composed of two groups of websites, where all those in the first group have links to all those in the second. Often, this motif identifies a group of 'fans' with the same interests (say, blogs on rafting), pointing to their 'idols' (say, websites of rafting magazines). Networks that regulate genes are almost completely built out of motifs. When the bacterium *E. coli* is in a stressful condition, a specific genetic circuit senses the stress and coordinates the production of certain proteins. These proteins coalesce to build flagella, a kind of moving tail that allows the bacterium to swim away in search of better conditions. This same genetic circuit, the *coherent feed-forward loop*, is present in many other bacteria and several other organisms. Evolution seems to have selected specific motifs, because of their optimal properties, (e.g. because they use the smaller number of genes necessary to perform a certain function). Moreover, a clear advantage of modularity is that motifs can combine to give rise to new functions, and damage to one of them is not propagated to the others.

Motifs are a kind of small-scale, local, repeated modules. But when people think about communities, they usually aim at finding great partitions of a network, such as compartments of a foodweb, online communities, disciplinary areas, etc. These structures do not show a regular, repetitive pattern. The task of finding them is easier if we have some clue, for example if the members of the community are self-identified by some element. This might be a widget all the members of the community add to their blogs, a common way of dressing, etc. However, most of the time this information is neither available nor explicit, and we have to dig into the network structure to find the modules. The general objective of community identification is to find sets of nodes that are more highly interconnected between themselves than with each other, such as in the karate club network. This verbal formulation is easy, but the mathematical translation of the concept is elusive, such that a definitive community detection method has not yet been found. Some methods aggregate nodes to fulfill an optimality criterion. Others split the network into groups, and then split the groups further, and then split them even more, to create a genealogic tree of nested communities for the partition. Others place imaginary springs between nodes and look at the clusters formed after the relaxation of the system. In general many other options are available. An interesting technique, that cleverly uses the topology of the network, is based on computing the *edge betweenness*. That is, finding the edges through which most of the shortest paths pass. The links with highest edge betweenness are akin to the weak links connecting otherwise separated groups in Granovetter's work. If one cuts a few edges with high betweenness, then the network splits into a certain number of isolated clusters: these are suitable candidate communities. One can continue cutting the higher-betweenness edges to find more detailed structures nested within the larger ones.

An interesting application of community finding is the analysis of the US political blogosphere. Physicist Lada Adamic found a clear separation between Democrats' and Republicans' blogs. The resulting structure showed two large groups with very few connections to each other. Moreover, the structure related to liberals' blogs was found to be less cohesive than the conservatives' one. For instance, in the part of this blogosphere dedicated to abortion, pro-life blogs show a denser interconnection than pro-choice blogs. As a result, an online campaign is likely to spread more easily in the first group than in the second. Another study analysed communities of students in US schools, to see whether ethnicity shaped social networks. In both very diverse and very homogeneous schools, ethnicity seemed to be irrelevant. On the contrary, segregation was visible for intermediate values of diversity. In the metabolic network, communities have been found to correspond to specific functions (carbohydrate metabolism, nucleotide and nucleic acid metabolism, protein, peptide, and amino acid metabolism, lipid metabolism, aromatic compound metabolism, monocarbon compound metabolism, and coenzyme metabolism). Finally, company stocks have been clustered on the basis of price correlations, finding modules corresponding to the various business areas, such as banking, mining, distribution, financial, etc.

The definition of community as a subgraph 'more densely connected internally than externally' is very general, but fails to capture some special modules. Think of a telephone chain between friends, in which the first calls the second, the second calls the third, and so on: such a chain would most probably not be classified as a community, according to this definition. Another example is the web pages of competitors working in the same business: obviously, they do not have incentives to link to each other, although they clearly belong to the same community. In addition, real-world communities are much more complex than dense clusters. Several partitions are possible at the same time; nationality, social class, gender, job, political ideas: all could classify the same set of people. Moreover, communities can overlap: an individual can belong to multiple nationalities or affiliations at the same time. Finally, nested communities are possible: for example, regional identities within national identities.

Albeit a strong simplification, the graph representation is still capable of capturing many relevant features of a system. A close look at a graph provides plenty of information and more details arise when more complex measurements are performed. Almost all the time, real-world networks deviate from their random counterparts, suggesting the existence of some kind of built-in order. Again, all these networks are not blueprinted: these deviations very likely arise from self-organization processes. Finding new regularities in graph structures and revealing the underlying mechanisms are some of the ongoing challenges of network science.

Chapter 8
Perfect storms in networks

Settings for surprise

The island of Barro Colorado is a piece of rainforest in the middle of the waters of the Panama Canal. When a nearby river was dammed, just a few hilltops remained uncovered. The island has become an open-air experiment about what happens to a forest when it is fragmented into small pieces, as when highways, buildings, fields or mines substitute the original vegetation. A few years after the inundation around Barro Colorado, the population jaguars and pumas had shrunk dramatically. As a consequence, their prey thrived: now, in the island there is plenty of specimens of a large rodent, called agouti. These animals love the big seeds of acacias, so their boom is a big problem for acacias to successfully reproduce, as well as for microorganisms colonizing their seeds. As the acacia population shrinks, plants producing smaller seeds occupy their place, and animals eating them also increase in number. The original alteration of the ecosystem extends in all directions in the foodweb of the island.

Domino effects are not uncommon in foodwebs. Networks in general provide the backdrop for large-scale, sudden, and surprising dynamics. Pathogens spreading in transport networks, blackouts in power grids, large conflicts, or unexpected cooperative efforts in social systems: networks seem to be the ideal setting for 'perfect storms'. Network nodes can represent individual entities (people, computers, species, genes . . .) exchanging material or information (information packets, energy, etc.), or they can represent locations (countries, airports . . .) exchanging individual entities (goods, travellers . . .). Within this very broad classification, the range of possible dynamics is enormous. Why are networks the natural playground for all these dynamics? How does the graph structure influence these processes? A general answer is impossible, but in many cases we can see that the heterogeneous, non-random organization of the underlying network makes a big difference to all the phenomena taking place on top of it.

Failures and attacks

On 18 July 2001, a train derailed in an underground tunnel in Baltimore (US), and began a fire. Soon after, the Internet was slowed down in several states along the US east coast. The fire had burnt optic cables passing through the tunnel, that belonged to several of the most important Internet Service Providers of the country. As a consequence, the accident created a domino effect that spanned a large part of the US. The Internet is constantly exposed to similar accidents. A percentage of routers are always out of operation at any time, for a broad range of reasons. Potentially, each one of these accidents may be as serious as Batimore's derailment. Still, such macroscopic damages are rare. The network seems to tolerate a certain amount of chronic dysfunction without too many problems. It relies somehow on alternative paths, allowing traffic to get around failures. Still, like most networks, the Internet does not have many redundant links, and is not highly dense either. With these features in mind, it would be natural to expect it to break down easily.

While the Internet seems to be relatively resistant to errors and accidents, a carefully designed attack can wreak terrible damage. On 7 February 2000, an enormous number of users logged on to the Yahoo! website. There were so many that the company servers were not able to answer these requests and the web page went down. In the days that followed, a set of other web pages, ranging from eBay to CNN, went down for the same reason. After two months, the police discovered that the logons were artificial and came from a 15-year-old Canadian hacker, whose nickname was Mafia Boy. He did not need to burn any cable to block the Internet: what he did was enough to bring down the websites that attracted most of the traffic on the WWW.

As with the Internet and the WWW, most of the real-world networks show a double-edged kind of robustness. They are able to function normally even when a large fraction of the network is damaged, but suddenly certain small failures, or targeted attacks, bring them down completely. For example, genetic mutations arise naturally throughout life (and some of them can even delete certain proteins from the cell) or are produced artificially (as in the case of a genetic technique called *gene knockout*, that turns off the function of a whole gene in lab rats). Still, organisms display a great tolerance to many mutations, and to an unexpectedly large number of knockouts. Most of the time they continue to work normally, in overall terms. On the other hand, certain specific mutations are capable of completely disrupting the workings of a cell. The brain loses neurons all the time: a stressful experience for any given organ, such as getting drunk occasionally, can kill a considerable number of cells. But after the hangover everything works fine again, usually. In Parkinson's disease, a large portion of the neurons can disappear without the patient even noticing. But when this portion exceeds a certain threshold, then the disruptive condition starts to become manifest.

In this respect, networks are very different from engineered systems. In an airplane, damaging one element is enough to stop the whole machine. In order to make it more resilient, we have to use strategies such as duplicating certain pieces of the plane: this makes it almost 100 per cent safe. In contrast, networks, which are mostly not blueprinted, display a natural resilience to a broad range of errors, but when certain elements fail, they collapse. How many errors can a network tolerate without even noticing the problem? And what are the elements that cause the collapse? With the objective of answering these questions, scientists have simulated failures by removing nodes from network maps and observing what happens. After the removal of a fraction of nodes, they check whether the surviving nodes are still connected (that is, whether a giant connected component is still present in the network) and close (that is, whether the average distance is still small). In order to simulate errors, the nodes are removed at random. When this is done to a random network, after a few removals the distance increases quickly and the graph breaks down in many disconnected components. A random graph of the size of most real-world networks is destroyed after the removal of half of the nodes. On the other hand, when the same procedure is performed on a heterogeneous network (either a map of a real network or a scale-free model of a similar size), the giant connected component resists even after removing more than 80 per cent of the nodes, and the distance within it is practically the same as at the beginning. The scene is different when researchers simulate a targeted attack, as in the strategy of Mafia Boy. They started by removing first the most 'important' nodes (hubs) of the network. In this situation the collapse happens much faster in both networks. However, now the most vulnerable is the second: while in the homogeneous network it is necessary to remove about one-fifth of its more connected nodes to destroy it, in the heterogeneous one this happens after removing the first few hubs.

Highly connected nodes seem to play a crucial role, in both errors and attacks. They are the 'Achilles heel' of most heterogeneous networks exposed to targeted attacks. In these networks, hubs are mainly responsible for the overall cohesion of the graph, and removing a few of them is enough to destroy it. On the other hand, hubs are also the 'ace in the hole' of these networks, when they are exposed to errors and failures: when nodes are removed at random, most of the time the selected nodes come out from the large population with low degree, so as long as hubs are kept untouched, the network stays together. This behaviour becomes clearer considering that the degree is usually correlated with the betweenness. High-degree nodes are most of the time bridges through which many paths of the network pass. When random damage is applied to a network, it will rarely affect one of the few hubs. While hubs are unaffected, they provide the necessary connectivity: there is no need for many redundant connections; paths crossing hubs keep the working areas of the damaged network connected. In those few networks in which some low-degree nodes have high betweenness and act like bridges (as certain airports do), attacking hubs still causes serious damage, but the most lethal strategy is attacking the most central nodes.

Domino effects

The possibility of a sudden transition from a resilient behaviour to a global collapse should ring some alarm bells. In ecosystems, a certain rate of extinction is inevitable: one in each million of species becomes extinct every year, according to some estimates. Usually, foodwebs rearrange after these events, and most of the species do not suffer major damage from these natural extinctions. But large-scale collapses are possible too: about 250 million years ago, more than 90 per cent of the species disappeared in a relatively short period, the famous Permian extinction. Five massive extinctions of this kind have been registered in the last 500 million years. Researchers have argued that external factors may be the cause, such as the much-debated meteorite that may have made dinosaurs extinct. However, a network explanation is also possible. Cases of extinctions in chain, or *co-extinctions*, are not unknown to ecologists. For example, the introduction of the virus of myxomatosis to control the population of rabbits in England in the mid-20th century ended up making the big blue butterfly (*Maculinea arion*) extinct in 1979. The virus decimated rabbits, and as a consequence the tall grass they ate spread in the fields. This destroyed the habitat of ants, that used to make nests in low grass, where the sun could reach. Ants had a mutualistic relation with blue butterflies' larvae: they took care of the larvae, which responded by providing liquid food to the ants. The disruption of their habitat gradually impaired the reproduction of the butterflies, bringing them to extinction. This is not a coextinction in the literal sense, since rabbits did not disappear due to mixomatosis and blue butterflies have been partially reintroduced. However, it gives an idea of how far damage to foodwebs can go. A large-scale version of this story, with a chain of extinctions that depletes almost a full ecosystem of species, is a possible alternative explanation to the great extinctions of the past. This should also be taken into account when massive attacks on ecosystems are voluntarily carried out by humans, as in the case of too much fishing currently depleting marine ecosystems at an unprecedented scale.

Several other dynamic processes on networks could give rise to similar *cascading failures*, or *breakdown avalanches*. A typical example is a large-scale blackout: the failure of a power station overloads another one, which fails in its turn, propagating the overload throughout a large part of the network. In this phenomenon, the failure of a node results not only in loss of interconnection or reduction of the average distance but also in a domino effect. The *systemic failure* of economic networks experienced during financial crises is another instance of this phenomenon. The same can happen in *congestion phenomena*, such as cars collapsing certain points of the street network, people collapsing a subway station during a special event, or online traffic collapsing certain Internet services. In all these cases, studies have shown that hubs are crucial, both because they reduce transit times and because they are first in becoming saturated.

Epidemics

In 1347, one of the most devastating plagues in human history appeared in Constantinople. During the following three years, the Black Death moved to Europe, leaving a large fraction of its population dead. The disease covered Europe like a wave, at a velocity of 200–400 miles per year (Figure 12 left). This picture is much different from that of modern pandemics. The 1918 influenza that is estimated to have killed 3 per cent of the world population took just one year to spread, reaching even isolated Pacific islands in that time. The 1957 flu virus, also called 'Asian flu', swept the globe in about six months. More recent outbreaks, such as 2009's swine flu, leapt from one side of the planet to the other in a few weeks (Figure 12 right). While the Black Death travelled with pilgrims, merchants, and sailors, lurking in ships and carriages, at a few miles per day, modern diseases can rely on much more efficient means of transportation, such as highways, trains, and aeroplanes. In the 14th century, physical distance was a leading factor in the spread of an epidemic. In the modern networked world, an infection can jump on a plane and reach the opposite side of the planet in a few hours.

Epidemics spread in networks both at the global level (for example, through the airport network) and at the local level: infectious diseases that jump from person to person depend on individuals' social networks. For example, flu spreads partially through face-to-face contact between individuals, while HIV spreads in the network of unprotected sexual contacts. In 2001, the Cabilan physicists Romualdo Pastor Satorras and his Italian colleague Alessandro Vespignani decided to study the problem by modelling and simulating the spread of a disease in a social network. They introduced just the minimal ingredients of an infectious process: at the beginning, a few individuals of a social network get infected; if a healthy individual is in contact with one of them through a

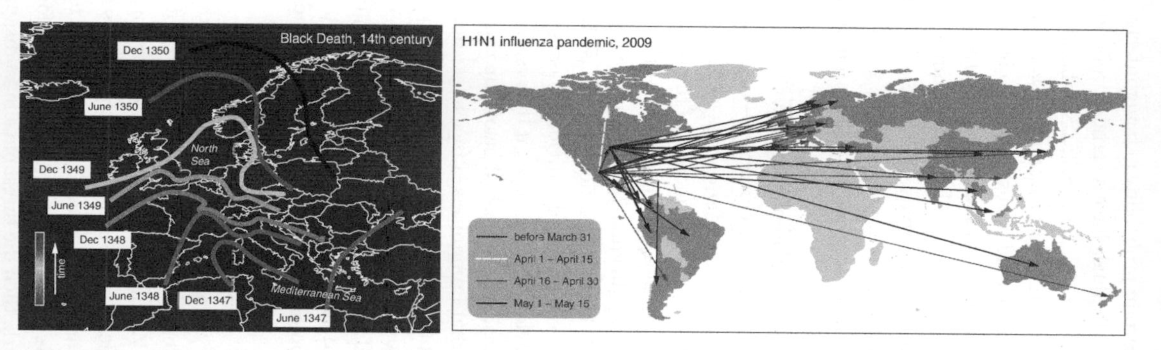

12. The 14th-century bubonic plague swept Europe like a wave (left), while 2009 swine flu was more similar to a fire throwing sparks from one side of the planet to another (right): the difference is due to the dramatic change in human transportation networks

link, he or she has a certain probability of being infected; on the other hand, infected individuals have a certain probability of recovering. This model of infection is called *SIS*, because each individual passes through the cycle *susceptible-infected-susceptible* (a healthy individual is 'susceptible' to being infected). The process represents infections such as the common cold, from which people usually recover. It can be further complicated to represent other diseases, for example by introducing the possibility that people die or are immunized. However, the general direction of the results are not changed by these modifications.

Pastor Satorras and Vespignani found that, after an initial phase of expansion, the virus can either be eradicated—it shrinks and finally disappears from the population—or become endemic—it sustains itself and infects a certain fraction of the population indefinitely. The disease is said to be below the *epidemic threshold* when, for every infected individual, fewer than one person gets infected: in this case, it becomes extinct. The disease is above the epidemic threshold if every infected individual passes the disease to more than one individual: in this case, it thrives. If vaccines are available, the disease can be pushed below the threshold by means of campaigns that immunize a sufficient portion of the population. Very contagious diseases are usually the hardest cases, because they have a low epidemic threshold and so they become endemic very easily. If eradication is too hard, pushing the disease closer to the threshold still has a positive effect: that is, reducing the proportion of people indefinitely affected by the endemic disease.

In their study, Pastor Satorras and Vespignani found that the epidemic threshold depends crucially on the features of the underlying network. When the SIS dynamics are performed on a random network, a clear threshold is found that allows us to estimate how many individuals have to be immunized to extinguish the disease. But when the dynamics are performed on a heterogeneous network, then the threshold almost disappears: it is much lower than in the random graph; moreover, the larger the size of the system, the lower the threshold. In a large enough network, the threshold is so low that it is almost unavoidable to have a proportion of infected individuals. The disease cannot be pushed below such a low threshold without immunizing almost all the population. In epidemics, as in many other dynamics, heterogeneity makes a difference. Studies of errors and attacks have shown that hubs keep different parts of a network connected. This implies that they also act as bridges for spreading diseases. Their numerous ties put them in contact with both infected and healthy individuals: so hubs become easily infected, and they infect other nodes easily. The *super-spreaders* identified by epidemiologists are likely the hubs of social networks.

The vulnerability of heterogeneous networks to epidemics is bad news, but understanding it can provide good ideas for containing diseases. Ideally, almost all the population should be immunized to block the infection completely. However, if we can immunize just a fraction, it is not a good idea to choose people at random. Most of the times, choosing at random implies selecting individuals with a relatively low number of connections. Even if they block the disease from spreading in their surroundings, hubs will always be there to put it back into circulation. A much better strategy would be to target hubs. Inmunizing hubs is like deleting them from the network, and the studies on targeted attacks show that eliminating a small fraction of hubs fragments the network: thus, the disease will be confined to a few isolated components. This strategy faces a practical problem: nobody really knows the full map of social connections of a human group, so it is hard to identify its hubs. However, a clever tactic to find them was suggested in 2003 by physicists Reuven Cohen, Shlomo Havlin, and Daniel ben-Avraham: they suggested selecting people at random and asking them the name of somebody they are connected with. The most repeated names in this list are most likely the hubs of the social network: in fact, for its many connections, a hub will be tied to many people, so it will probably be mentioned by many of those interviewed. It should be noted that immunizing hubs works perfectly in theory, but many real-world details could impair it, such as whether the network disposes of specially redundant paths that go around hubs, or whether the network of contacts is fixed in time or evolving: for example, if Alice has HIV, and has unprotected sex with Bob, and Bob has unprotected sex with Carol, it makes a big difference to Carol whether Bob has sex with her before or after having sex with Alice.

The picture of the spread of an epidemic in a social network can be partially generalized to the case in which nodes do not represent people but locations (say, airports), and what spreads on the network are people (say, infected or healthy travellers). In this case, one can define a *global invasion threshold*, above which the disease becomes a pandemic, and below which it remains contained at local level. Closing airports is rarely a good idea: we would need to shut down 90 per cent of airports to block certain epidemics effectively, which would have too high a social and economic cost. Cleverer strategies, such as sharing antivirals with developing countries (which are often the source of new pandemics), are much more effective.

Viruses, ads, and fads

A couple of obscure Pakistani programmers, a university professor, a group of high school students . . . these were the authors of the first computer viruses. During the eighties, these parasite programs started to jump from one computer to another, basically hiding in the floppy disks interchanged by users. The first viruses were academic experiments on self-replicating software, but soon they escaped from the lab. In 1986, the *Brain* virus appeared from Pakistan. In the same year, a German laboratory lost control of *Virdem*. One year later, a group of students put *Vienna* in circulation. In the nineties, computer viruses were already a global problem, but this was nothing compared to what was in store.

The advent of the Internet brought a new generation of viruses that were capable of sending themselves to other computers through the Net. In 1999, *Melissa* spread through the Internet: people started to receive email messages with subjects such as 'Important message for you' or 'Here is the document you asked for... don't show anyone else ;-)'. The mail contained a file called list.doc. If the receiver opened it, it launched a program that sent the same message to the first fifty email addresses held in the computer. *Iloveyou*, *Slammer*, *Sobig*, *Blaster*, and many others exploded across the Net, using similar mechanisms and with disastrous effects: some of them destroyed companies' computer systems, universities' databases, and even affected Internet traffic.

Some features of a computer virus infection are strikingly similar to real-world epidemics. A computer becomes infected through its connections (for example, the social contacts of its owner as sampled by his or her email network) and infects others similarly. Some of the conclusions reached for diseases explain the puzzling behaviour of computer viruses. Even if antivirus programs are quickly updated, some viruses still circulate years after their first strike. This is no surprise if one considers the features of an epidemic spreading in scale-free networks: even if a large proportion of computers are immunized through antivirus programs, this is not enough to eradicate the infection: there is always some high-degree node here or there that puts it back in circulation.

This characteristic of endurance, which is a real problem with computer viruses, can be turned into a resource if one wants to disseminate information in a heterogeneous network. This is the principle behind *viral marketing*. Thanks to virtual social networks, today the WWW is full of videos, games, and applications that have 'gone viral': they are being forwarded by hundreds of thousands of people to all their contacts every day. One of the first examples of this idea was the spread of the Hotmail email service. In 1996, the company inserted into emails an automatic footer saying 'Get your free web-based email at

Hotmail', containing a link to a form for setting up a new email address in a few seconds (see page 50). Similar strategies were implemented by the email services of Yahoo! and Google, and by many social networks that are launched on the basis of providing access by invitation only.

Viral marketing takes advantage of an underlying psychological phenomenon called *social spreading*. This is the general tendency of people to mimic their contacts' behaviours, and to spread gossip, fads, rumours and ideas. This mechanism also acts in innovation adoption, group problem solving, and collective decision making. Sociologists and psychologists have found many examples of the striking tendency of humans to 'copy' each other. In 1962, a group of girls at a mission school in Tanzania experienced an unusual tendency to uncontrollable laughter. After a few months, tens of pupils of the school showed the same symptoms, and other people in the villages where some of the pupils were sent to rest showed the same disquieting giggling. After much investigation, doctors A. M. Rankin and P. J. Philip, who studied the case, came to the conclusion that it was an instance of 'mass hysteria'. A similar case was reported in 1998 in a high school in Tennessee, when the experience of a teacher who had the feeling of smelling gasoline spread to hundreds of students. All environmental factors were excluded, and scientists came to the conclusion that a kind of 'emotional contagion' was at work.

Many similar cases of social spreading have been documented, but in recent years scientists have found that the same mechanism may play a role in less exceptional settings: for example, obesity and smoking seem to spread on social networks. Three reasons are behind the fact that people connected share certain features or behaviours. First, there are external factors, such as belonging to the same social class: for example, people belonging to lower social classes have an increased risk of smoking and becoming obese; at the same time they are more likely to establish ties with each other than with people of a higher social class. Second, there is homophily: people that smoke or have similar body mass tend to make friends with those with similar habits. Third, there is social spreading: if you are a friend of somebody who smokes or is overweight, you are more likely to consider taking up smoking or increasing your daily food intake. Probably all three mechanisms are at work, but social spreading is likely to be the least trivial of them and should not be underestimated. Sociologists argue that it's not a specific condition that spreads; rather, it's the sharing of norms about what is appropriate that is disseminated. This perspective could be used in public health, to foster safer habits by targeting hubs in social networks.

Naturally, the contagion of behaviours, as well as rumours and ideas, is different in many respects from that of diseases. Unlike contagion, the act of spreading information is necessarily intentional. On the other hand, acquiring information is usually advantageous, so it is a more active process than getting infected. Learning or being convinced may need a longer exposure than getting a disease. Moreover, many other competing mechanisms are present. If social spreading was the leading factor, uniformity would be the rule, but in fact mechanisms against simple assimilation generate diversity, minorities, and polarization. In any case, in certain settings social contagion may indeed be the most relevant mechanism. In the forties, Richard Feynman invented *Feynmann's diagrams* as tools for modern high-energy physics. Some physicists accepted them with enthusiasm and others with scepticism, but they finally triumphed. A study of their diffusion in the communities of physicists of the US, Japan and USSR revealed that the observed trends could be quite accurately fitted with models used for epidemics, provided that the parameters were tuned to very different values.

Which came first, networks or dynamics?

One of the keys to the success of ancient Rome was its strategic position close to the River Tiber, which at the time was first and foremost a communication and commercial route. When the city became more powerful, it started building the first branches of its formidable road network. In their turn, the roads were a crucial tool for maintaining and further expanding Rome's power, since they provided a quick way to move goods and legions. More roads meant more power, and more power made it necessary to create even more roads: the result was that 'All roads lead to Rome,' according to an Italian saying. Similar patterns of development can be observed in almost every important city. A growing city attracts traffic and requires more connections (roads, railways, airlines . . .), which in their turn increase traffic and growth, which imply even more connections, etc. The communication network influences the dynamics of traffic, but this in turn reshapes the network, in a feedback loop.

Asking how network topology affects dynamics implies an assumption: that the network is an immutable structure, on top of which processes take place. In reality, all networks change *during* the dynamics. Therefore this assumption makes sense only if the timescale of the dynamics is much faster than that of the topology. This is reasonable for certain processes: for example, information interchanged on a daily or weekly basis spreads on a fixed social network, since usually friendship and kinship have turnover times in the order of years; or vehicle traffic towards a city during any given day moves on a fixed set of pathways: usually the street connections do not change every day.

However, in other cases this assumption is flawed. For example, in the epidemic spread of sexually transmitted diseases the timing of the links is crucial. Establishing an unprotected link with a person before they establish an unprotected link with another person who is infected is not the same as doing so afterwards. If we want to study the development of a city throughout a decade, it is necessary to take into account the interplay between traffic and changing connections. In certain technological networks, such as peer-to-peer file-sharing systems, network structure and information dynamics change on the same timescale and are strongly interwoven. In foodwebs, population dynamics can produce a reorganization of the network. When overfishing pushes a species below a certain level, the foodweb rearranges the predations and new species take the place of the old one. The coupling of network structure and dynamics is especially relevant at a time of virtual social networking. These tools provide a constant information flow about the structure and content of a person's social network. So researchers argue that this enhanced awareness may alter the way in which people create, maintain, and leverage their social networks.

Several approaches are possible to cope with the problem of coupling network structure and dynamics. For example, one can build network models through optimization, in which the quantity to be optimized is related to dynamics such as flow or searching. A more refined approach consists of modifying the fitness model in such a way that the value of the fitness depends on some dynamic parameter. When dynamics proceed, fitness changes accordingly; and this allows a reshaping of the networks. Other strategies are possible, all of them underpinning a basic idea: when a dynamic takes place in or is coupled to a network structure, then most of the time it is essential to take into account the underlying graph to fully understand what is going on.

Chapter 9

All the world's a net; or not?

One of the fathers of quantum mechanics, Paul Dirac, was reported to say about the revolutionary discoveries of physicists at the beginning of the 20th century: 'the rest is chemistry'. He meant that all science could be derived from the first principles of physics. Unfortunately, no more than a few cases, essentially the atoms of hydrogen and helium, can be solved precisely with quantum mechanical equations. More complex objects, like molecules, must be approached through approximations or computer simulations. Apart from a few macroscopic quantum effects, at the moment fundamental physics is relatively useless in understanding biology, the mind, or society. Similar mistakes occur in genetics, when DNA is incorrectly framed as something that can explain all the features, diseases, and behaviours of humans. In general, the results of basic science should not be taken beyond their real range of effectiveness, and it should be acknowledged that more specialized disciplines can give much deeper insights beyond that range. Network science should avoid the trap of overhyping. Its holistic vision, the revelation of unexpected similarities between widely different systems, and the current cultural fascination with the concept of network bring with them the temptation to think of network science as a 'theory of everything'. Sociologists, engineers, biologists, and philosophers have warned about the bald generalizations drawn from network theory. Most of this criticism is reasonable, but the results of network science should not be underestimated, nor its potential for future discoveries impaired.

The first major limitation of network science is the hunger for large-scale data. Methods used in social science, such as questionnaires and interviews, are costly, time consuming, and sometimes prone to subjective biases. Data drawn from information technologies (phone calls, email, social networks, geo-localization, RFID chips, health data, credit cards, etc.) provide unprecedented access to people's social relations, but they pose some problems too. A person delivering pizzas receives many phone calls, but most of them are from clients, not from friends: the *pizza delivery guy problem* shows that sorting out relevant information from large amounts of IT data (the list of phone calls, in this example) is not easy. Moreover, it should be mentioned that data mining with networks also creates ethical problems related to privacy and to their use by the military.

In many situations, only partial data are available. In order to draw aquatic foodwebs, ecologists capture fish and examine their digestive tracts: with this method, even the most brilliant scientist can miss some of the links. Genetic methods to deduce physical interactions between proteins can produce both false positives and false negatives. Maps of the Internet and the WWW are obtained by sending a 'probe' from a node to explore the surrounding edges: a sufficient number of paths can give a fairly good representation of the network, but some edges may never be discovered by the probes.

Once the data are available, representing them in a graph is inevitably a reductionist act. The focus on topology is one of the strengths of the network approach, but it forgets many of the specific features of the elements. If we are interested in those features, the graph approximation can be inadequate. Sometimes, but not always, network models can be modified to include these features.

The graph representation can fall short when geography (that is, the physical location of nodes) is more important than topology. For example, the position of electrical substations, airports, or train stations is obviously relevant to the arrangement of their connections. Furthermore, in social networks and foodwebs, closeness can determine the actual possibility of establishing a relation. Another element that may escape graph representations is time. For example, in sexually transmitted diseases, establishing a link with a person *after* he or she is infected by another person is obviously very different from doing so *before* the infection: the timing of links is crucial for the spread of the disease. Timing is important in a large range of networks: for example, scientific publications are arrayed in time, and they can only cite papers that came out in the past.

Sometimes, identifying nodes and edges is not trivial. It is easy to distinguish between an eagle and a hawk, but one could easily lose count of the number of bacteria in an ecosystem. To avoid underestimating the number of 'small' species, ecologists customarily aggregate organisms in *trophic species*, sets of organisms that share the same predators and the same preys. Similarly, social scientists aggregate individuals that are *structurally equivalent*: people that have the same number and kind of ties, for example within a family group. Similar procedures are applied to the Internet at the autonomous system level, the network of brain areas, etc. These procedures have to be performed in a coherent way to obtain a network that makes sense. Defining edges can be even more complicated. A company can hold a small fraction of the capital of another one, or up to 100 per cent of it. Two airports can be connected by one flight per day or by one per hour. In all these cases, a threshold must be established, below which a relation is considered too weak to be recorded. Weighting or putting a threshold on links strongly influences the shape of the resulting network, and must be done with very good reasons.

Once data have been arranged as a graph, a careful interpretation of the results is essential. An entire branch of network science is devoted to visualization, that is, producing algorithms to arrange nodes and links sensibly on paper or computer screen. However, most of the conclusions cannot be drawn by visual inspection, and mathematical analysis is needed. Criticisms have been made on the basis that not all complex networks have a heterogeneous degree distribution (and in any case this is never a mathematically exact power law). It is true that networks can be interesting also if they are not heterogeneous, but it is fair to say that often interesting networks are indeed heterogeneous. Perfect power laws are not vital: what matters is the presence of fat tails, revealing the existence of hubs. Interpreting heterogeneity as a signal of self-organization has been criticized, pointing out that a certain level of planning is present in many networks, as in the case of the precise design of local networks by administrators in the Internet. However, there is no doubt that the Internet, like many other networks, has not been designed at the large-scale level. So it is reasonable to argue that their deviations from randomness may be attributed to self-organization. Moreover, degree heterogeneity is just one signal of complexity in networks. Heterogeneity is also found in other features, such as their betweenness, clustering, weights, etc., and complexity is visible as well in other features, different from heterogeneity: for example, the modularity and community structure, which often deviates from randomness and has strong effects on the dynamics that take place on top of networks.

Another criticism points to the fact that network science has found nothing more than fuzzy similarities between different systems and dynamics, but not real *universality classes*. These are groups of different phenomena that correspond to the same basic mathematical laws (once certain details are discounted). Certainly, the specific features of a biological network are totally different from those of a technological one, and the diffusion of a computer worm has different rules from the contagion of a disease. However, network theory provides a framework of shared trends and common predictions for such different structures and processes. Usually, if the system is big enough and the phenomena are observed for long enough, they will display fairly similar trends.

Network science has shown predictive power in several fields. It is currently being used in consulting, to help organizations to better exploit the skills distributed across its members; in public health, to predict and prevent the spread of infectious diseases; in police and the military, to track terrorist, criminal, or rebel networks; and in several other fields. There are many problems to be tackled: among them, making more detailed models to fit specific networks and dynamics; finding new relevant network data; digging deeper in topology to find unnoticed regularities and fully explain the existing ones; characterizing small networks and learning how to deal with networks of networks; connecting biological networks more effectively with the evolutionary paradigm; finding new applications (for example in drug design); and possibly finding universality classes.

A famous quote from Galileo Galilei states: 'Philosophy is written in this grand book, the universe . . . it is written in the language of mathematics, and its characters are triangles, circles, and other geometric figures . . . .' We believe that, especially in our complex contemporary world, we now need the 'characters' of networks.

Further reading

Popular science books on networks

Albert-László Barabási, *Linked, The New Science of Networks*. Perseus, New York (2002)

Mark Buchanan, *Nexus: Small Worlds and the New Science of Networks*. W. W. Norton & Co, New York (2002)

Ricard Solé, *Redes complejas. Del genoma a Internet*. Tusquets, Barcelona (2009)

Nicholas Christakis, James Fowler, *Connected. The Surprising Power of Our Social Networks and How They Shape Our Lives*. Little, Brown, New York (2009)

Introductory academic books on networks

Guido Caldarelli, *Scale-Free Networks*. Oxford University Press, Oxford (2007)

Romualdo Pastor-Satorras, Alessandro Vespignani, *Evolution and Structure of the Internet. A Statistical Physics Approach*. Cambridge University Press, Cambridge (2004)

Alain Barrat, Marc Barthelemy, Alessandro Vespignani, *Dynamical Processes on Complex Networks*. Cambridge University Press, Cambridge (2008)

L. C. Freeman, *The Development of Social Network Analysis: A Study in the Sociology of Science*. Empirical Press, Vancouver (2004)

Stanley Wasserman, Katherine Faust, *Social Networks Analysis. Methods and Applications*. Cambridge University Press, Cambridge (1994)

Albert-László Barabási et al. Interactive Network Science textbook project http://barabasilab.nev.edu/networksciencebook/ (in progress)

Articles and reviews

Mark Newman, Albert-László Barabási, Duncan J. Watts, *The Structure and Dynamics of Networks*. Princeton University Press, Princeton (2006) (includes bibliographic essays and an anthology of seminal papers in network theory)

Réka Albert, Albert-László Barabási, Statistical mechanics of complex networks. *Review of Modern Physics*, 74, 47–97 (2002)

M. E. J. Newman, The Structure and Function of Complex Networks. *SIAM Review*, 45 (2), 167–256 (2003)

S. Boccaletti et al. Complex Networks: Structure and Dynamics. *Physics Reports*, 424 (4–5), 175–308 (2006)

Katy Börner, Soma Sanyal, Alessandro Vespignani, Network Science. *Annual Review of Information Science and Technology*, 41 (1), 537–607 (2007)

Stephen P. Borgatti et al. Network Analysis in the Social Sciences. *Science*, 323 (5916), 892–95 (2009)

Index

C

D

E

I

J

K

L

M

N

Expand your collection of
VERY SHORT INTRODUCTIONS

1. Classics
2. Music
3. Buddhism
4. Literary Theory
5. Hinduism
6. Psychology
7. Islam
8. Politics
9. Theology
10. Archaeology
11. Judaism
12. Sociology
13. The Koran
14. The Bible
15. Social and Cultural Anthropology
16. History
17. Roman Britain
18. The Anglo-Saxon Age
19. Medieval Britain
20. The Tudors
21. Stuart Britain
22. Eighteenth-Century Britain
23. Nineteenth-Century Britain
24. Twentieth-Century Britain
25. Heidegger
26. Ancient Philosophy
27. Socrates
28. Marx
29. Logic
30. Descartes
31. Machiavelli
32. Aristotle
33. Hume
34. Nietzsche
35. Darwin
36. The European Union
37. Gandhi
38. Augustine
39. Intelligence
40. Jung
41. Buddha
42. Paul
43. Continental Philosophy
44. Galileo
45. Freud
46. Wittgenstein
47. Indian Philosophy
48. Rousseau
49. Hegel
50. Kant
51. Cosmology
52. Drugs
53. Russian Literature
54. The French Revolution
55. Philosophy
56. Barthes
57. Animal Rights
58. Kierkegaard
59. Russell
60. Shakespeare
61. Clausewitz

62. Schopenhauer
63. The Russian Revolution
64. Hobbes
65. World Music
66. Mathematics
67. Philosophy of Science
68. Cryptography
69. Quantum Theory
70. Spinoza
71. Choice Theory
72. Architecture
73. Poststructuralism
74. Postmodernism
75. Democracy
76. Empire
77. Fascism
78. Terrorism
79. Plato
80. Ethics
81. Emotion
82. Northern Ireland
83. Art Theory
84. Locke
85. Modern Ireland
86. Globalization
87. The Cold War
88. The History of Astronomy
89. Schizophrenia
90. The Earth
91. Engels
92. British Politics
93. Linguistics
94. The Celts
95. Ideology
96. Prehistory
97. Political Philosophy
98. Postcolonialism
99. Atheism
100. Evolution
101. Molecules
102. Art History
103. Presocratic Philosophy
104. The Elements
105. Dada and Surrealism
106. Egyptian Myth
107. Christian Art
108. Capitalism
109. Particle Physics
110. Free Will
111. Myth
112. Ancient Egypt
113. Hieroglyphs
114. Medical Ethics
115. Kafka
116. Anarchism
117. Ancient Warfare
118. Global Warming
119. Christianity
120. Modern Art
121. Consciousness
122. Foucault
123. The Spanish Civil War
124. The Marquis de Sade
125. Habermas
126. Socialism
127. Dreaming
128. Dinosaurs
129. Renaissance Art
130. Buddhist Ethics
131. Tragedy
132. Sikhism
133. The History of Time

134. Nationalism
135. The World Trade Organization
136. Design
137. The Vikings
138. Fossils
139. Journalism
140. The Crusades
141. Feminism
142. Human Evolution
143. The Dead Sea Scrolls
144. The Brain
145. Global Catastrophes
146. Contemporary Art
147. Philosophy of Law
148. The Renaissance
149. Anglicanism
150. The Roman Empire
151. Photography
152. Psychiatry
153. Existentialism
154. The First World War
155. Fundamentalism
156. Economics
157. International Migration
158. Newton
159. Chaos
160. African History
161. Racism
162. Kabbalah
163. Human Rights
164. International Relations
165. The American Presidency
166. The Great Depression and The New Deal
167. Classical Mythology
168. The New Testament as Literature
169. American Political Parties and Elections
170. Bestsellers
171. Geopolitics
172. Antisemitism
173. Game Theory
174. HIV/AIDS
175. Documentary Film
176. Modern China
177. The Quakers
178. German Literature
179. Nuclear Weapons
180. Law
181. The Old Testament
182. Galaxies
183. Mormonism
184. Religion in America
185. Geography
186. The Meaning of Life
187. Sexuality
188. Nelson Mandela
189. Science and Religion
190. Relativity
191. The History of Medicine
192. Citizenship
193. The History of Life
194. Memory
195. Autism
196. Statistics
197. Scotland
198. Catholicism
199. The United Nations
200. Free Speech
201. The Apocryphal Gospels

202. Modern Japan
203. Lincoln
204. Superconductivity
205. Nothing
206. Biography
207. The Soviet Union
208. Writing and Script
209. Communism
210. Fashion
211. Forensic Science
212. Puritanism
213. The Reformation
214. Thomas Aquinas
215. Deserts
216. The Norman Conquest
217. Biblical Archaeology
218. The Reagan Revolution
219. The Book of Mormon
220. Islamic History
221. Privacy
222. Neoliberalism
223. Progressivism
224. Epidemiology
225. Information
226. The Laws of Thermodynamics
227. Innovation
228. Witchcraft
229. The New Testament
230. French Literature
231. Film Music
232. Druids
233. German Philosophy
234. Advertising
235. Forensic Psychology
236. Modernism
237. Leadership
238. Christian Ethics
239. Tocqueville
240. Landscapes and Geomorphology
241. Spanish Literature
242. Diplomacy
243. North American Indians
244. The U.S. Congress
245. Romanticism
246. Utopianism
247. The Blues
248. Keynes
249. English Literature
250. Agnosticism
251. Aristocracy
252. Martin Luther
253. Michael Faraday
254. Planets
255. Pentecostalism
256. Humanism
257. Folk Music
258. Late Antiquity
259. Genius
260. Numbers
261. Muhammad
262. Beauty
263. Critical Theory
264. Organizations
265. Early Music
266. The Scientific Revolution
267. Cancer
268. Nuclear Power
269. Paganism
270. Risk
271. Science Fiction

272. Herodotus
273. Conscience
274. American Immigration
275. Jesus
276. Viruses
277. Protestantism
278. Derrida
279. Madness
280. Developmental Biology
281. Dictionaries
282. Global Economic History
283. Multiculturalism
284. Environmental Economics
285. The Cell
286. Ancient Greece
287. Angels
288. Children's Literature
289. The Periodic Table
290. Modern France
291. Reality
292. The Computer
293. The Animal Kingdom
294. Colonial Latin American Literature
295. Sleep
296. The Aztecs
297. The Cultural Revolution
298. Modern Latin American Literature
299. Magic
300. Film
301. The Conquistadors
302. Chinese Literature
303. Stem Cells
304. Italian Literature
305. The History of Mathematics
306. The U.S. Supreme Court
307. Plague
308. Russian History
309. Engineering
310. Probability
311. Rivers
312. Plants
313. Anaesthesia
314. The Mongols
315. The Devil
316. Objectivity
317. Magnetism
318. Anxiety
319. Australia
320. Languages
321. Magna Carta
322. Stars
323. The Antarctic
324. Radioactivity
325. Trust
326. Metaphysics
327. The Roman Republic
328. Borders
329. The Gothic
330. Robotics
331. Civil Engineering
332. The Orchestra
333. Governance
334. American History
335. Networks